Algorithms for Variable-Size Optimization

Applications in Space Systems and Renewable Energy

Ossama Abdelkhalik

Associate Professor
Department of Aerospace Engineering
Iowa State University

CRC Press
Taylor & Francis Group
Boca Raton London New York

CRC Press is an imprint of the
Taylor & Francis Group, an **informa** business

A SCIENCE PUBLISHERS BOOK

First edition published 2021
by CRC Press
6000 Broken Sound Parkway NW, Suite 300, Boca Raton, FL 33487-2742

and by CRC Press
2 Park Square, Milton Park, Abingdon, Oxon, OX14 4RN

CRC Press is an imprint of Taylor & Francis Group, LLC

ISBN: 9780815360162 (hbk)

Typeset in Times New Roman
by Radiant Productions

For my kids Jumana, Minna, Nour, and Ganna

Preface

Optimization methods have evolved significantly over the past several decades. It is a combination of the fast growing computational capabilities, and the bigger challenges engineers face when designing more aggressive systems for the prosperity of human beings, which push for the development of more capable optimization algorithms.

The core of this book presents recent advances in global optimization algorithms. The focus is on *The Hidden Genes Genetic Algorithms* and *The Structured Chromosome Evolutionary Algorithms*. The motivation for developing these methods was to develop algorithms for interplanetary trajectory design optimization. This problem in particular is characterized by the fact that the number of design variables is itself a variable. One candidate solution to a trajectory optimization problem of a given space mission may have a number of optimization variables that are different from that of another candidate solution for the same mission. Another way to describe that, which is adopted here in this book, is that the size of the design space is variable. In fact, the interplanetary trajectory design optimization is not unlike several other system architecture optimization applications, characterized also by a variable-size design space, and hence these methods may be applicable in several other applications.

This book is suitable for graduate students in engineering, engineers, and researchers. The book collects research published in several recent publications by the author, and represents it in a one coherent text. The first chapter of the book presents a brief review for mathematical fundamentals, along with examples of several engineering applications in which the size of the design space is variable. The second chapter presents detailed mathematical modelling developments for three engineering design applications; these three examples are the design of Earth orbits, the design of interplanetary trajectories, and the design and control of ocean wave energy converters. Some background on the fundamentals of Astrodynamics is needed to fully follow the modeling work presented in the first two examples. These three examples are used in later chapters as example applications for The Hidden Genes Genetic Algorithms and The Structured Chromosome Evolutionary Algorithms. **Section** II presents a summary for most common classical optimization methods used for unconstrained and constrained optimization problems. **Section** III presents the The Hidden Genes Genetic Algorithms, The Structured Chromosome Evolutionary Algorithms, and The Dynamic-Size Multiple Population Genetic Algorithms. The three methods presented in **Section** III are designed to search for optimal solutions in variable-size design space optimization problems. **Section** IV of the book presents detailed applications in interplanetary trajectory design optimization, and in ocean wave energy conversion.

The algorithms and results of The Hidden Genes Genetic Algorithms presented in this book are based upon work supported by the National Science Foundation under Grant Number 1446622. The methods presented in **Section** III were developed based on foundations laid down by giants in this field. In particular I was influenced by the work of Bruce A. Conway, Massimiliano Vasile, and Dario Izzo. **Section** II was written with influence and inspiration from several recent textbooks; in particular I would like to thank P. Venkataraman for his contribution in presenting classical optimization methods in a concise yet clear way for engineers and engineering students. I also thank my students who improved the material presented in this book over the years.

April 2020   **Ossama Abdelkhalik**

Contents

Preface   **iv**

Symbol Description   **x**

SECTION I: BACKGROUND AND MOTIVATION

1. Introduction and Background   **2**

    1.1 Mathematical Background   3
        1.1.1 Definitions   3
        1.1.2 Differentiability and Taylor's Theorem   4
        1.1.3 Orthogonal Vectors   5
                1.1.3.1 Gram-Schmidt Orthogonalization   6
                1.1.3.2 Q-Orthogonal (Q-Conjugate) Directions   7
        1.1.4 Convergence Rates   7
    1.2 Systems Architecture Optimization   8
        1.2.1 Interplanetary Trajectory Optimization   8
        1.2.2 Microgrid Optimization   11
        1.2.3 Traffic Network Signal Coordination Planning   11
        1.2.4 Optimal Grouping Problems   12
        1.2.5 Systems Design Optimization   12
        1.2.6 Structural Topology Optimization   13
        1.2.7 Pixel Classification Problems   14

2. Modeling Examples of Variable-Size Design Space Problems   **15**

    2.1 Satellite Orbit Design Optimization   15
    2.2 Interplanetary Trajectory Optimization   19
    2.3 Optimization of Wave Energy Converters   25

SECTION II: CLASSICAL OPTIMIZATION ALGORITHMS

3. Fundamentals and Core Algorithms — 34

   3.1 Equal Interval Search Algorithm — 35
   3.2 Golden Section Method — 37
   3.3 Linear versus Nonlinear Optimization — 41
      3.3.1 Linear Programming — 42
      3.3.2 Nonlinear Programming — 43

4. Unconstrained Optimization — 47

   4.1 Non-Gradient Algorithms — 47
      4.1.1 Cyclic Coordinated Descent Method — 48
      4.1.2 Pattern Search Method — 48
      4.1.3 Powell's Method — 51
   4.2 Gradient-Based Optimization — 51
      4.2.1 Steepest Descent Method — 52
      4.2.2 Conjugate Gradient Method — 52
      4.2.3 Variable Metric Methods — 56
   4.3 Second Order Methods — 58

5. Numerical Algorithms for Constrained Optimization — 60

   5.1 Indirect Methods — 61
      5.1.1 Barrier Methods — 61
      5.1.2 Exterior Penalty Function Methods — 63
      5.1.3 Augmented Lagrange Multiplier Method — 65
      5.1.4 Algorithm for Indirect Methods — 67
   5.2 Direct Methods — 68
      5.2.1 Sequential Linear Programming — 68
      5.2.2 Quadratic Programming — 69
      5.2.3 Sequential Quadratic Programming — 72

SECTION III: VARIABLE-SIZE DESIGN SPACE OPTIMIZATION

6. Hidden Genes Genetic Algorithms — 76

   6.1 Introduction to Global Optimization — 76
   6.2 Genetic Algorithms — 78
      6.2.1 Similarity Templates (schemata) — 80
      6.2.2 Markov Chain Model — 80
   6.3 Fundamental Concepts of Hidden Genes Genetic Algorithms — 84
      6.3.1 The Hidden Genes Concept in Biology — 84

      6.3.2 Concept of Optimization using Hidden Genes Genetic   85
           Algorithms
      6.3.3 Outline of a Simple HGGA   86
  6.4 The Schema Theorem and the Simple HGGA   89
      6.4.1 Reproduction   89
      6.4.2 Crossover   93
      6.4.3 Mutation   93
  6.5 Hidden Genes Assignment Methods   94
      6.5.1 Logical Evolution of Tags   95
      6.5.2 Stochastic Evolution of Tags   95
  6.6 Examples: VSDS Mathematical Functions   97
      6.6.1 Examples using Stochastically Evolving Tags   98
      6.6.2 Examples using Logically Evolving Tags   104
  6.7 Statistical Analysis   104
  6.8 Markov Chain Model of HGGA   107
  6.9 Final Remarks   115

7. Structured Chromosome Genetic Algorithms   117

  7.1 Structured-Chromosome Evolutionary Algorithms (SCEAs)   118
      7.1.1 Crossover in SCGA   120
      7.1.2 Mutation in SCGA   121
      7.1.3 Transformation in SCDE   122
      7.1.4 Niching in SCGA and SCDE   123
  7.2 Trajectory Optimization using SCEA   124
      7.2.1 Earth-Mars Mission   126
      7.2.2 Earth-Saturn Mission (Cassini 2-like Mission)   128
      7.2.3 Jupiter Europa Orbiter Mission   131
  7.3 Comparisons and Discussion   136

8. Dynamic-Size Multiple Population Genetic Algorithms   140

  8.1 The Concept of DSMPGA   140
  8.2 Application: Space Trajectory Optimization   142
  8.3 Numerical Examples   145
  8.4 Discussion   152

SECTION IV: APPLICATIONS

9. Space Trajectory Optimization   156

  9.1 Background   156
  9.2 A Simple Implementation of HGGA   160
      9.2.1 Optimization   160

	9.2.2	Numerical Results	164
	9.2.3	Discussion	172
9.3	Trajectory Optimization using HGGA with Binary Tags		176
	9.3.1	Earth–Jupiter Mission using HGGA	176
	9.3.2	Earth–Jupiter Mission: Numerical Results and Comparisons	177

10. Control and Shape Optimization of Wave Energy Converters   **183**

10.1 A Conical Buoy in Regular Wave   187
10.2 General Shape Buoys in Regular Waves   189
10.3 WECs in Irregular Waves   194
     10.3.1 Simultaneous Optimization of Shape and Control   197
10.4 Discussion   198

Bibliography   **204**

Index   **215**

Symbol Description

T	time of flight, days	$\Delta\mathbf{v}_{ps}$	powered swing-by impulse, km/s
$\Delta\mathbf{V}_d$	departure impulse, km/s		
$\Delta\mathbf{V}_a$	arrival impulse, km/s	η	swing-by plane rotation angle, rad
$\Delta\mathbf{V}_{DSM}$	deep space maneuver impulse, km/s	Π	perpendicular plane to the incoming relative velocity
$\Delta\mathbf{V}$	mission total cost, km/s	Γ	intersection line between Π and the inertial ecliptic plane
n	number of impulses in N-impulse maneuver		
ε	epoch of a DSM as a fraction of transfer time	Ω	angle between Γ and the inertial $\hat{I}$, rad
$\mathbf{r}$	heliocentric position vector in inertial frame, km	i	inclination of Π to the ecliptic, rad
$\mathbf{v}_\infty$	spacecraft relative hyperbolic velocity, km/s	C	transformation matrix
		P_i	planet identification number
$\mathbf{v}_p$	planet velocity vector, km/s	F_i	fitness (cost function)
$\mathbf{v}_\infty^-$	incoming relative velocity, km/s	N_s	number of swing-by maneuvers
$\mathbf{v}_\infty^+$	outgoing relative velocity, km/s	D_i	number of DSM in each mission leg
$\mathbf{r}_{per}$	pericenter radius vector, km	F_d	flight direction
δ	deflection angle in the swing-by plane, rad	t_d	departure date
		t_a	arrival date
μ_p	swing-by planet gravitational constant, km³/s²	$\bar{h}_i$	normalized swing-by pericenter altitude

L_{max}	maximum chromosome length	T	leg time of flight, *days*
i	maximum possible number of swing-by maneuvers	$\mathbf{v}$	heliocentric velocity vector in inertial frame, *km/s*
j	maximum possible number of total DSMs in the whole trajectory	$\vec{x}$	vector of variables
		x_j^*	*j*th variable of the optimal solution
k	maximum number of independent thrust impulses	μ	gravitational constant, *km³/s²*
q	number of swing-bys followed by a zero-DSM trajectory	η	flyby plane rotation angle, *rad*
		ε	DSM epoch (fraction of T)
$\mathbf{C}$	transformation matrix	Π	perpendicular plane to incoming relative velocity
C	child solution		
f	fitness function	ι	inclination of Π to ecliptic, *rad*
g_i	*i*th gene in the chromosome	Γ	intersection line between Π and inertial ecliptic plane
g_{pj}^i	*i*th gene in the parent *j*		
h	pericenter altitude, *km*	Ω	angle between Γ and inertial $\hat{I}$, *rad*
J	fitness value	λ	random variable between 0 and 1
M	number of tags		
m	number of flybys	a	arrival
N	number of genes	d	departure
n	number of deep-space maneuvers	*DSM*	Deep Space Maneuver
		mod	modified
P	planet ID (1(Mercury) to 8(Neptune))	*npf*	non-powered flyby
		p	planet
P_{ti}	gene string of parent solution *i*	*pf*	powered flyby
		S/C	spacecraft
$\mathbf{r}$	heliocentric position vector in inertial frame, *km*	*tot*	total
		+	outgoing
r_{per}	pericenter radius of flyby, *km*	−	incoming
t	Julian date		

BACKGROUND AND MOTIVATION

I

Chapter 1

Introduction and Background

Optimization is a critical step in a wide range of engineering disciplines including structural design optimization, topology optimization, space trajectory planning, product designs, proxy models optimization, and the inverse problem optimization found in several applications. Engineering Optimization - sometimes referred to as Design Optimization - utilizes optimization methods and algorithms to achieve desired design goals in engineering. The engineer or designer uses mathematical optimization to support decision making or improve a given design. In particular, real-valued function minimization is one of the most common optimization problems, where a parameter space is searched for the minimum value of the objective function, and the corresponding parameters values.

In searching for an optimal design, a first step is to mathematically formulate the engineering problem. An optimization problem usually has the following mathematical format:

$$\begin{aligned}
\text{minimize: } & f(\vec{x}) \\
\text{subject to: } & g_i(\vec{x}) \le b_i, \quad i = 1 \cdots m, \\
& h_j(\vec{x}) = 0, \quad j = 1 \cdots l,
\end{aligned}$$
$$\text{and the boundaries: } \vec{x}^l \le \vec{x} \le \vec{x}^u \tag{1.1}$$

The function $f : \mathbb{R}^n \to \mathbb{R}$ is a scalar objective function that is constructed based on the goal of optimization. The vector $\vec{x} = [x_1, \cdots, x_n]^T$ contains the n design variables which the designer can vary to achieve the optimization goal. The functions $g_i : \mathbb{R}^n \to \mathbb{R}, i = 1, \cdots, m$, are the inequality constraint functions, which are set up based on the problem constraints and limits. The scalars $b_i, i = 1, \cdots, m$, are the limits which are also derived from the problem constraints. The functions $h_j : \mathbb{R}^n \to \mathbb{R}, j = 1, \cdots, l$, are the equality constraint functions. The set of all the feasible points of $\vec{x}$ (feasible solutions) is called the design space.

Once the optimization problem is formulated in the form of (1.1), the engineer then needs to decide on how to solve this specific mathematical optimization problem. Solving this problem means finding the optimal value $\vec{x}^*$ for the variable vector $\vec{x}$ that will make the function $f(\vec{x}^*)$, minimum among all the feasible vectors $\vec{x}$. The solution method or algorithm depends on several factors. There are few classes of optimization problems, and for each class there are specific methods that can solve optimization problems in this class. Optimization problems may be classified based on the existence of constraints, the nature of the design variables, and the nature of the equations involved. For example, if all the functions f, $h_j, \forall j$, and $g_i, \forall i$ are linear then this optimization problem is called a linear programming problem. Linear programming problems are natural in operations research. There are methods that can solve this type of problem very effectively such as simplex methods. If any of the objective or the constraint functions is nonlinear, then this problem is called nonlinear programming. There are a few special cases that can be solved very effectively such as least-squares problems. Otherwise, in general, there is no effective manner for solving the general nonlinear programming problem without compromise.

Engineering optimization then involves the formulation of the engineering problem into a mathematical optimization problem, and then solving this problem. There could be multiple mathematical formulations for the same engineering problem, and different formulations may have different levels of difficulty in terms of solving the problem. The problem formulation is then a crucial step and this book presents case studies on how to formulate an engineering optimization problem.

1.1 Mathematical Background

1.1.1 Definitions

A vector $\vec{x} \in \mathbb{R}^n$ has n scalar components, $\vec{x} = \begin{bmatrix} x_1 \\ x_2 \\ \vdots \\ x_n \end{bmatrix} = \begin{bmatrix} x_1 & x_2 & \ldots x_n \end{bmatrix}^T$, where the

superscript T is the transpose.

The norm of a vector ($\|.\|$) is a real-valued function $\|\vec{x}\| : \mathbb{R}^n \to \mathbb{R}$. There are various norms:

$l_1 - norm : \|\vec{x}\|_1 = \sum_{i=1}^{n} |x_i|$

$l_2 - norm : \|\vec{x}\|_2 = (\sum_{i=1}^{n} |x_i|^2)^{\frac{1}{2}}$ (distance or length of $\vec{x}$)

$l_p - norm:$ $(\sum_{i=1}^{n} |x_i|^p)^{\frac{1}{p}}$

$l_\infty - norm:$ $\|\vec{x}\|_\infty = Max_i\{|x_i|\}$

The norm of a vector satisfies the following properties:

1. $\|\vec{x}\| \geq 0, \forall \vec{x}$,

2. $\|\vec{x}\| = 0$ if and only if $\vec{x} = \vec{0}$,

3. $\|\vec{x} + \vec{y}\| \leq \|\vec{x}\| + \|\vec{y}\|, \forall \vec{x}$ and $\vec{y}$

1.1.2 Differentiability and Taylor's Theorem

The real-valued function $f(x) : \mathbb{R} \rightarrow \mathbb{R}$ is said to be differentiable at point a if the limit

$$\lim_{h \rightarrow 0} \frac{f(a+h) - f(a)}{h} \quad (1.2)$$

exists. If the limit exists, it is called "derivative" of f at a, and is denoted $f'(a)$. The function f is a "differentiable" function if it is differentiable at each point in its domain.

For functions of multiple variables, the function $f(\vec{x}) : \mathbb{R}^n \rightarrow \mathbb{R}$ is said to be differentiable at point $\vec{a}$ if there exists a linear transformation $L : \mathbb{R}^n \rightarrow \mathbb{R}$ that satisfies the condition

$$\lim_{\vec{h} \rightarrow 0} \frac{f(\vec{a} + \vec{h}) - f(\vec{a}) - L(\vec{h})}{\|\vec{h}\|} = 0. \quad (1.3)$$

The vector associated with the linear transformation L is the vector of partial derivatives; that is $L(\vec{h}) = \nabla f(\vec{a}) \cdot \vec{h}$.

Let $f(x)$ be an a function of a single variable, and let $f(x)$ be infinitely differentiable in some open interval around $x = x_0$. The Taylor series of $f(x)$ around x_0 is

$$f(x_0 + \Delta x) = \sum_{k=0}^{\infty} \frac{f^k(x_0)}{k!} (\Delta x)^k \quad (1.4)$$

Hence, a quadratic approximation for $f(x)$ would be

$$f(x_0 + \Delta x) \approx f(x_0) + f'(x_0)\Delta x + \frac{f''(x_0)}{2!} (\Delta x)^2 \quad (1.5)$$

For functions of multiple variables, assume a function $f(\vec{x}) : \mathbb{R}^n \rightarrow \mathbb{R}$ is infinitely differentiable in some open interval around $\vec{x} = \vec{x}_0$. The Taylor series expansion can be written as

$$f(\vec{x}_0 + \Delta \vec{x}) = f(\vec{x}_0) + \nabla f(x_0) \cdot \Delta \vec{x} + \Delta \vec{x} \cdot (H(\vec{x}_0)\Delta \vec{x}) + \cdots \quad (1.6)$$

where H is the matrix of second derivatives, called the Hessian matrix. For the case when $n = 2$, the Hessian matrix is

$$H(\vec{x}) = \begin{bmatrix} f_{11}(\vec{x}) & f_{12}(\vec{x}) \\ f_{21}(\vec{x}) & f_{22}(\vec{x}) \end{bmatrix} \tag{1.7}$$

where f_{ij} is the order partial derivative with respect to the variable $\vec{x}(i)$ and $\vec{x}(j)$.

The quadratic approximation for the function $f(\vec{x})$ is

$$f(\vec{x}_0 + \Delta\vec{x}) \approx f(\vec{x}_0) + \nabla f(x_0) \cdot \Delta\vec{x} + \Delta\vec{x} \cdot (H(\vec{x}_0)\Delta\vec{x}) \tag{1.8}$$

1.1.3 Orthogonal Vectors

For any two vectors $\vec{x}$ and $\vec{y}$ in the n-dimensional Euclidian space $\mathbb{R}^n$, the scalar (inner) product is defined as:

$$(\vec{x}, \vec{y}) = \vec{x} \cdot \vec{y} = \vec{x}^T \vec{y} = \sum_{i=1}^{n} x_i y_i \tag{1.9}$$

where each of the x_i and y_i is the ith component of the vectors $\vec{x}$ and $\vec{y}$, respectively, and $\vec{x}^T$ is the transpose of the vector $\vec{x}$.

The projection of a vector $\vec{x}$ on a vector $\vec{y}$ is defined as:

$$Proj_{\vec{y}}\vec{x} = \frac{(\vec{x}, \vec{y})}{\|\vec{y}\|_2^2}\vec{y} \tag{1.10}$$

Two vectors are said to be orthogonal if they are perpendicular to each other; that is is the scalar product of the two vectors is zero. The vectors $\{\vec{x}^1, \vec{x}^2, \cdots, \vec{x}^m\}$ are said to be mutually orthogonal if every pair of vectors is orthogonal. A set of vectors is said to be orthonormal if every vector is of unit magnitude and the set of vectors are mutually orthogonal. A set of vectors that are orthogonal are also linearly independent.

Consider for example the two vectors $\vec{x}^T = [8 \times 10^{-10}, 1.2 \times 10^6]$ and $\vec{y}^T = [2 \times 10^{-9}, 1.5 \times 10^5]$ (these vectors could be measurements of some process.) If we try to measure the size of each vector using the Euclidian norm, then we get $\|\vec{x}\|_2 \simeq |\vec{x}(2)|$ and $\|\vec{y}\|_2 \simeq |\vec{y}(2)|$; this is because the first component is very small compared to the second component. This means we loose information if we use the Euclidean norm. If we select a diagonally weighted norm $\|\vec{x}\|_Q = \sqrt{\vec{x}^T[Q]\vec{x}}$, where $[Q]$ is a diagonal matrix with positive elements, chosen to normalize each entry, then we will get a better representation. For instance if we let $[Q] = \begin{bmatrix} 10^{18} & 0 \\ 0 & 10^{-12} \end{bmatrix}$ then $[Q]^{\frac{1}{2}}\vec{x} = \begin{bmatrix} 8 \\ 1.2 \end{bmatrix}$ and $[Q]^{\frac{1}{2}}\vec{y} = \begin{bmatrix} 2 \\ 1.5 \end{bmatrix}$, in which the two components are comparable. Note that:

$$\|[Q]^{\frac{1}{2}}\vec{x}\|_2 = \sqrt{\left(\vec{x}^T[Q]^{\frac{T}{2}}\right)[Q]^{\frac{1}{2}}\vec{x}} = \sqrt{\vec{x}^T[Q]\vec{x}} = \|\vec{x}\|_Q \tag{1.11}$$

The general form for the scalar product is known as the Hermitian form, which is defined as:

$$(\vec{x},\vec{y})_Q = \vec{y}^{\dagger}[Q]\vec{x} \tag{1.12}$$

where $\vec{y}^{\dagger}$ is the conjugate transpose of $\vec{y}$ and $[Q]$ is any Hermitian positive definite matrix. A Hermitian matrix is a complex square matrix that is equal to its own conjugate transpose. In the case of real numbers, this scalar product can be thought of as the scalar product of the directionally scaled vectors, using positive scaling. So, if $\vec{x}$ and $\vec{y}$ are two vectors in $\mathbb{R}^n$ and $[Q]$ is symmetric positive definite $n \times n$ matrix, then $(\vec{x},\vec{y})_Q = \vec{x}^T[Q]\vec{y}$ is a scalar product. If $(\vec{x},\vec{y})_Q = 0$ then the vectors $\vec{x}$ and $\vec{y}$ are said to be Q-orthogonal.

1.1.3.1 Gram-Schmidt Orthogonalization

Orthogonalization is the process of finding a set of orthogonal vectors that span a particular subspace. This section presents a systematic process that can be used to construct a set of n orthogonal vectors given a set of n linearly independent vectors in $\mathbb{R}^n$. The process can be illustrated by a simple example. Consider two linearly independent vectors $\vec{x}^1 = \begin{bmatrix}01\end{bmatrix}^T$ and $\vec{x}^2 = \begin{bmatrix}11\end{bmatrix}^T$. These two vectors are not orthogonal. It is possible to construct two orthogonal vectors, however, given $\vec{x}^1$ and $\vec{x}^2$ as follows. Let the two orthogonal vectors be $\vec{y}^1$ and $\vec{y}^2$. We start the process by selecting $\vec{y}^1 = \vec{x}^1$. To compute $\vec{y}^2$, we follow this process:

$$\vec{y}^2 = \vec{x}^2 - Proj_{\vec{x}^1}\vec{x}^2 = \begin{bmatrix}1\\0\end{bmatrix} \tag{1.13}$$

$$(\vec{y}^2,\vec{y}^1) = 0 \tag{1.14}$$

The two vectors $\vec{y}^1$ and $\vec{y}^2$ are orthogonal. This process can be generalized for n vectors as follows. Given a set of linearly independent vectors $\vec{x}^1,\vec{x}^2,\cdots,\vec{x}^n$ in $\mathbb{R}^n$. Then we can construct a set of orthogonal n vectors using the process:

$$\vec{y}^1 = \vec{x}^1$$
$$\vec{y}^2 = \vec{x}^2 - Proj_{\vec{y}^1}\vec{x}^2$$
$$\vdots \tag{1.15}$$
$$\vec{y}^k = \vec{x}^k - \sum_{i=1}^{k-1} Proj_{\vec{y}^i}\vec{x}^k, \quad k \le n$$

Figure 1.1 illustrates the geometrical meaning of this process in three-dimensional space.

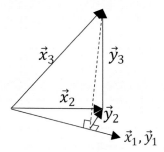

Figure 1.1: Gram-Schmidt Orthogonalization in 3-dimenional space.

1.1.3.2 Q-Orthogonal (Q-Conjugate) Directions

Similar to the development presented in **Section** 1.1.3.1, it is possible to construct a set of Q-orthogonal (Q-conjugate) n directions given a set of linearly independent n vectors $\vec{x}^1, \vec{x}^2, \cdots, \vec{x}^n$ in $\mathbb{R}^n$ using the following process:

$$\vec{d}^1 = \vec{x}^1$$

$$\vec{d}^2 = \vec{x}^2 - \frac{\left(\vec{x}^2, \vec{d}^1\right)_Q}{\|\vec{d}^1\|_Q^2} \vec{d}^1$$

$$\vdots \tag{1.16}$$

$$\vec{d}^k = \vec{x}^k - \sum_{i=1}^{k-1} \frac{\left(\vec{x}^k, \vec{d}^i\right)_Q}{\|\vec{d}^i\|_Q^2} \vec{d}^i, \; k \leq n$$

1.1.4 Convergence Rates

In most of the numerical optimization methods that are presented in this book, the search for the optimal solution is carried out iteratively. That is, at each iteration k, the algorithm finds a current guess (iterate) $\vec{x}^k$ for the optimal solution. The optimal solution is denoted as $\vec{x}^*$. Different algorithms converge to the optimal solution at different rates. Let $(\vec{x}^k)$ be a sequence of iterates converging to a local optimal solution $\vec{x}^*$. The sequence is said to converge:

Linearly: if there exists c, where $0 < c < 1$, and $k_{max} \geq 0$ such that $\forall k \geq k_{max}$:

$$\|\vec{x}^{k+1} - \vec{x}^*\| \leq c\|\vec{x}^k - \vec{x}^*\| \tag{1.17}$$

Super linearly: if there exists a null sequence c_k (a sequence converging to zero) of positive numbers, and $k_{max} \geq 0$, such that $\forall k \geq k_{max}$:

$$\|\vec{x}^{k+1} - \vec{x}^*\| \leq c_k\|\vec{x}^k - \vec{x}^*\| \tag{1.18}$$

Quadratically: if there exists $c > 0$ and $k_{max} \geq 0$, such that $\forall k \geq k_{max}$:

$$\|\vec{x}^{k+1} - \vec{x}^*\| \leq c\|\vec{x}^k - \vec{x}^*\|^2 \tag{1.19}$$

1.2 Systems Architecture Optimization

In many engineering systems, the system design optimization is crucial to achieve design objectives. The task of design optimization includes optimizing the system architecture (topology) in addition to the system variables. Optimizing the system architecture renders the dimension of the design space a variable (the number of design variables to be optimized is a variable.) This section presents few examples from engineering applications in which the optimization problem has a Variable-Size Design Space (VSDS). Hence, the optimization problem defined in Eq. (1.1), in the case of a VSDS problem, is modified to be:

$$\begin{aligned}
&\text{minimize: } f(\vec{x}, n) \\
&\text{subject to: } g_i(\vec{x}) \leq b_i, \quad i = 1 \cdots m. \\
&\qquad\qquad h_j(\vec{x}) = 0, \quad j = 1 \cdots l.
\end{aligned} \tag{1.20}$$

where n is the number of variables in $\vec{x}$. This type of problem is sometimes referred to as variable length problem. Most existing optimization algorithms can handle only Fixed-Size Design Space (FSDS) problems, and hence they cannot be used directly for system architecture optimization. The FSDS methods can be used to optimize the system variables, given a known system architecture. **Part** III presents recently developed algorithms that can handle VSDS optimization problems. Few examples of engineering applications are discussed in the following subsections to highlight the VSDS characteristic of systems architecture optimization problems.

1.2.1 Interplanetary Trajectory Optimization

Space trajectory design means calculating the history of the position vector $\vec{r}(t)$, of a spacecraft, from the launch date till the end date of its trip (e.g., arrival at the destination planet or asteroid). The dynamic model is a vectorial second order differential equation, that has no closed form solution [24]. To minimize the mission cost, a mission trajectory may turn out to be composed of several segments. The number of segments by itself is an unknown design variable. Consider a space mission to send a spacecraft from Earth to a planet in the solar system, with minimum fuel expenditure. The optimal trajectory could include fly-by other planets. A fly-by around a planet provides a free momentum change to the spacecraft. In a fly-by, the spacecraft becomes affected by the planet's gravitational field.

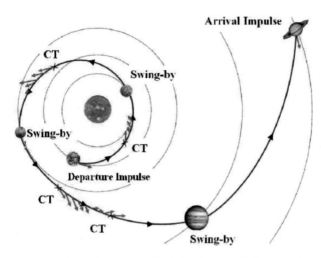

Figure 1.2: A space mission trajectory may include planetary fly-bys, continuous thrust segments, and Deep Space Maneuvers (DSMs).

Since planets orbit the Sun, the planets may be thought of as dynamic gravitational fields which a spacecraft may benefit from to minimize the trajectory cost. The trajectory may also include segments of continuous thrust firing. The thrust magnitudes and directions over time are design variables to be optimized. A Deep Space Maneuver (DSM) is another type of maneuver that uses large instantaneous impulses of thrust to change the spacecraft velocity instantaneously. Therefore, the design variables in this problem are: the number of fly-bys, the planets around which fly-bys will be carried out and the dates of fly-bys, the number of continuous thrust segments, the locations of these segments and the amount/directions of the thrust during these segments, the amount/direction of the launch impulse, and the launch and arrival dates. This optimization problem, in its general form, is challenging. On one hand, it includes both continuous variables and discrete variables (e.g., number of fly-bys, the fly-by planets, flight direction.) On the other hand, the objective function is replete with numerous local minima. In addition, the problem design space is not fixed: the number of design variables in a solution that consists of one segment is different from the number of design variables in another solution (probably with less cost) which consists of two segments.

One way to obtain a solution for this problem is to make few assumptions in order to convert it from a VSDS to a FSDS optimization problem. In such cases several optimization algorithms can be used such as the algorithms published in [66, 38]. For instance, a simplified version of the Cassini 2 mission trajectory design problem can be addressed by assuming a fixed fly-by sequence (EVVEJS) to be known a priori (E=Earth, V=Venus, J=Jupiter, and S=Saturn). **Figure** 1.3 shows the trajectory for the simplified Cassini 2 mission in such case.

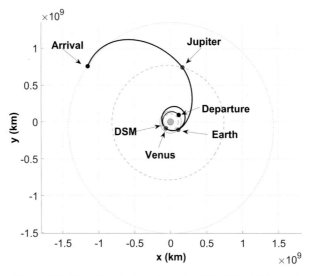

Figure 1.3: Overall trajectory for a simplified Cassini 2 mission.

Figure 1.4 zooms in the first part of the trajectory to show more details. **Part** III presents solutions to this problem obtained using VSDS optimization algorithms, and there is no need to make a priori assumptions on the fly-by sequence.

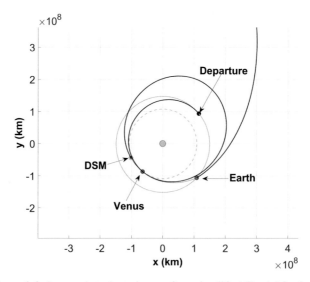

Figure 1.4: A zoom into the trajectory for a simplified Cassini 2 mission.

1.2.2   Microgrid Optimization

Consider a microgrid system where there are several energy sources and co-located energy storage devices that can either sink or source power with their corresponding sources. The net power at each source/storage is metered to the grid main bus using a boost converter. For an efficient design of the microgrid, the number of storage elements (N) and their capacities need to be optimized. Storage is expensive and designing a microgrid, with storage sized properly, is an open problem. Associated with computing the optimal N is the optimal values for the duty ratios at the converters that controls the power metered to the main bus from each source. A more complex situation is when we have M microgrids that have the ability to interconnect. This is a Variable-Size Design Space (VSDS) optimization problem that has a large number of permutations for exchanging power.

1.2.3   Traffic Network Signal Coordination Planning

Traffic Network Signal Coordination Planning (TNSCP) is usually formulated as an optimization problem [17, 61, 53, 54], and it is another application that motivates the need for VSDS optimization algorithms. The objective of this problem could be to maximize the traffic flow in one corridor (road), have higher overall vehicle speeds, reduce the number of stops, or some other objective. Most current methods for solving the TNSCP formulate the problem such that the signals' green times are the design variables [95], which is a formulation of fixed-size design space. This is a complex optimization problem, and extensive research has been conducted in using different optimization methods for this formulation, with different objective functions [78, 27, 102, 113, 107, 110, 80]. Usually, the green times values obtained from optimization are such that groups of signals share common values for the green times. **Figure** 1.5 is an illustration for this observation. The network in **Fig.** 1.5 has 20 intersections; some signals on the main streams at some intersections may have the same green time value. Hence it is possible to formulate the TNSCP optimization problem as an optimal grouping problem where it is desired to find the optimal grouping for all the optimized signals. Specific input values for the green times define the groups (subsets). The TNSCP optimization problem is then to determine which signals should belong to which group in order to optimize the objective function. The advantage of this formulation is that signals will be *coordinated* (using green times), to optimize the given objective, rather than treated as individual variables. In this traffic problem, it is the coordination between signals that is more effective than the green times of individual signals. This coordination information becomes out of context when using the standard formulation, while in the suggested formulation it is more direct. The number of variables (number of groups and number of signals in each group) is variable and needs a VSDS optimization tool.

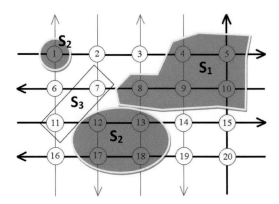

Figure 1.5: The TNSCP optimization problem is formulated as an optimal grouping problem where each group (subset $S_i, \forall i$) has a uniform green time value for all optimized signals in the group.

1.2.4 Optimal Grouping Problems

Optimal grouping problems have received a great deal of interest. Examples include the optimal bin packing problems [44], where a set of boxes of different sizes should be packed in containers of a given size, in order to minimize the number of containers. Reference [45] points out that standard FSDS global optimization algorithms are inefficient in handling this type of problem. A Grouping Genetic Algorithm (GGA) was developed to handle this type of problem, see, e.g., reference [43]. The GGA is group oriented, and hence is characterized by a VSDS (one solution may have two groups while another may have three groups). Tailored new definitions for genetic operations had to be introduced in the GGA theory in order to handle the resulting VSDS optimization problem.

1.2.5 Systems Design Optimization

Several systems design optimization problems are essentially VSDS problems. Here the thermodynamic cycle design optimization example is briefed. Thermodynamic cycle optimization is challenging as it involves several optimization variables and objectives. The first challenge (based on the choice of heat source, sink and working fluid) is to decide on the cycle topology or architecture. For example, in an exhaust waste heat recovery cycle that is applied as the bottoming cycle on a Gas Turbine (GT), the simplest cycle topology that can be applied is a Rankine cycle that utilizes a heater, a cooler/condenser, a compressor/pump and an expander (which is the separate turbine in **Fig.** 1.6). However, depending on the working fluid and the source/sink conditions, this basic cycle topology is not necessarily the best for achieving the most efficient cycle. Certainly, optimization of a given basic cycle configuration can be carried out with allowed inclu-

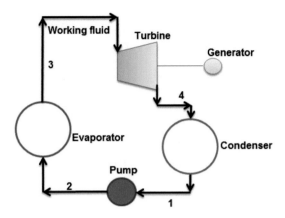

Figure 1.6: Schematic diagram of the Rankine cycle.

sions of intermediate cycles and components at appropriate locations (in terms of end-point pressures and temperatures). However, this is only part of the overall optimization needed and is typically straightforward. The broader and more challenging type of optimization is to consider several different types of allowed inclusions (of intermediate cycles and components) in the basic cycle configuration towards optimizing one or more well defined objective functions (such as overall power plant efficiency, cost, etc.) As discussed, the number of optimization variables in this broader optimization prospective changes depending on the particular cycle configuration and number of inclusions under consideration. In other words, in optimizing a given cycle, one solution may be completely specified with fewer variables whereas a range of valid choices may be available with larger number of variables.

1.2.6   Structural Topology Optimization

In structural topology optimization, the design objective could be to minimize the mass while satisfying a set of requirements in terms of the loads that a structure can handle. The design variables selection may vary depending on the problem formulation. Structural topology optimization problems, in general, are characterized by complex design spaces due to the large number of design variables. Standard global optimization methods become inefficient in such problems. For example, consider **Fig.** 1.7 where an example of three different designs for a structural elements is shown. If it is desired to optimize some objective, what would be optimization variables when using a FSDS algorithm? Is it possible to use a VSDS optimization algorithm?

Figure 1.7: Three different designs of a structural element.

1.2.7   Pixel Classification Problems

An important problem in satellite imagery is the classification of pixels for partitioning different landcover regions [23]. Satellite images usually have a large number of classes with nonlinear and overlapping boundaries. This problem has been formulated in the literature as a global optimization problem [35, 81], and several attempts have been made to solve it using GAs [35]. Reference [23], in particular, presents a method that minimizes the number of misclassified points, by assuming two possible sizes for the design space, and implementing a GA technique. The population in this case has two classes of chromosomes (males and females in [23] and new definitions for genetic operations had to be defined to handle the double-size design space. Clearly this type of problem would benefit from a VSDS optimization algorithm.

Chapter 2

Modeling Examples of Variable-Size Design Space Problems

The mathematical modeling of an engineering optimization problem is crucial for effective optimization. Modelling here means to mathematically express the objective and the constraint functions. These mathematical expressions depend on the physics of the problem and on how the problem is formulated. The problem formulation usually refers to the selection of the design variables and the objective function.

A design optimization problem may have multiple mathematical models that can be used. The model dictates the optimization methods that can solve the problem. The efficiency and/or the effectiveness of a solution method to a problem, in general, could be affected by the mathematical model being used. This chapter presents some case studies for engineering optimization problems, highlighting the thought process of formulating the problem and developing the mathematical models. Some of these case studies will be further addressed when discussing the system architecture optimization methods in subsequent chapters.

2.1 Satellite Orbit Design Optimization

In Earth orbiting space missions, the orbit selection dictates the mission parameters such as the ground resolution, the area coverage, and the frequency of coverage parameters. Earths regions of interest are identified, and one way for a

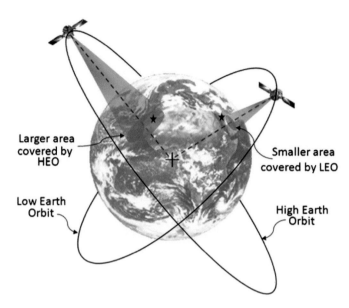

Figure 2.1: Spacecraft at high altitudes have larger coverage on Earth as compared to those at lower altitudes. The spacecraft speed at higher altitudes, however, is less than those at lower altitudes.

spacecraft to cover these regions is to continuously maneuver between them. This method is expensive: it requires a propulsion system onboard the space-craft, working throughout the mission lifetime. It also requires a longer time to cover all the regions of interest, due to the very weak thrust forces compared to that of the Earth's gravitational field. A natural orbit, in which the regions of interest are visited depending only on the gravitational forces without the use of a propulsion system, would be more economic and would need a short time period for visiting all sites. Lower altitude orbits enable a spacecraft to provide higher resolution measurements, but lack wide coverage and orbital perturbations are non-negligible due to atmospheric drag.

The problem can be formulated as an optimization problem. A penalty function is constructed to quantify the error of a given orbit from the ideal case of visiting all the sites within a given time frame. This penalty function is minimized to find the orbit for a given set of sites and a given time frame.

Problem Formulation

Assume we have n sites to be visited. Each site is defined by its geodetic longitude and latitude, λ_k and ϕ_k, respectively, where $k = 1, \ldots, n$. The difference between geodetic and geocentric latitudes is usually very small, and is neglected in this analysis. The position vector for the kth site in the Earth Centered Inertial

(ECI) frame is:

$$\mathbf{r}_k^I = R_E \left\{ \begin{array}{c} \cos\phi_k \cos(\lambda_k + \omega_E t) \\ \cos\phi_k \sin(\lambda_k + \omega_E t) \\ \sin\phi_k \end{array} \right\} \tag{2.1}$$

The only variable in the position vector of the kth site is the time, t, at which this site will be visited.

A space orbit can be completely specified using five orbital parameters; these are: the semi-major axis a, the eccentricity e, the inclination i, the argument of perigee ω, and the right ascension of the ascending node Ω. The position of the spacecraft on this orbit is determined using one additional angle, called the true anomaly φ. The spacecraft position vector can be computed at any location on the orbit using the previous six orbit parameters as briefed here. The satellite position $\mathbf{r}^o$ can be expressed in the perifocal coordinate system as:

$$\mathbf{r}^O = \frac{p}{1 + e\cos\varphi} \left\{ \begin{array}{c} \cos\varphi \\ \sin\varphi \\ 0 \end{array} \right\}, \tag{2.2}$$

where $p = h^2/\mu$ is the orbit parameter, h is the specific angular momentum of the spacecraft, and μ is a constant gravitational parameter. This vector is transformed to the ECI coordinate system through the transformation matrix, $R^{I/O}$ ($C_x \equiv \cos x$ and $S_x \equiv \sin x$):

$$R^{I/O} = \begin{bmatrix} C_\omega C_\Omega - C_i S_\omega S_\Omega & -S_\omega C_\Omega - C_i S_\Omega C_\omega & S_i S_\Omega \\ C_\omega S_\Omega + C_i S_\omega C_\Omega & -S_\omega S_\Omega + C_i C_\Omega C_\omega & -S_i C_\Omega \\ S_\omega S_i & C_\omega S_i & C_i \end{bmatrix} \tag{2.3}$$

by

$$\mathbf{r}^I = R^{I/O} \mathbf{r}^O \tag{2.4}$$

The time at which a ground site is visited (in Eq. (6.3)) is coupled with the spacecraft true anomaly (in Eq. (6.12)) through the Kepler equation.

$$\frac{\mu^2}{h^3} \left(1 - e^2\right)^{3/2} t = E - e\sin E, \tag{2.5}$$

where

$$\tan\left(\frac{E}{2}\right) = \sqrt{\frac{1-e}{1+e}} \tan\left(\frac{\varphi}{2}\right) \tag{2.6}$$

Consider a satellite with an observing instrument (radar, camera, etc.) with an aperture of ϑ_{FOV}. Two possible objectives are considered here: one is to maximize the resolution, and one is to maximize the observation time. For the best resolution case, a candidate optimality criterion is to minimize a weighted sum of squares of the distances between each site and the satellite at the nearest ground track point. The objective function

$$L_R = \sum_k \alpha_k (\mathbf{r}_k^I - \mathbf{r}^I)^T (\mathbf{r}_k^I - \mathbf{r}^I) \tag{2.7}$$

will drive a solution orbit to pass as near as possible to each site and also have the best achievable resolution since the resolution is proportional to $\|(\mathbf{r}_k^I - \mathbf{r}^I)\|$. For the observation time, the objective function is then expressed as:

$$L_T = \sum_k \alpha_k H(\frac{1}{2}\vartheta_{\text{FOV}} - \delta_k) \left(t_f - \int_0^{t_f} \cos\eta_k\, dt \right), \tag{2.8}$$

where η_k is the angle between $\mathbf{r}$ and $\mathbf{r}_k$ (see **Fig.** 2.2). $H(x)$ is the Heaviside unit step function $[H(x) = 0$ if $x < 0$, $H(x) = 1$ if $x > 0]$, and δ_k is the nadir angle, measured at the satellite from the nadir to the site.

The goal is to minimize the objective function L which is a function of a state vector (design variables to be optimized) whose elements are the orbital parameters a, e, i, ω, Ω, and all the visiting times t_k. A visiting time t_k is the time of closest approach to the site k. The minimizing state vector dictates the solution orbit. Reference [9] presents the detailed problem formulation for this problem as well as solution results when solving this problem. Reference [7] also presents the problem formulation for a more complex problem where the mathematical model takes into account the size of the field of view of the spacecraft sensor and the perturbation on the spacecraft motion due to the Earth oblateness.

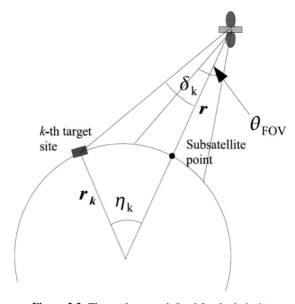

Figure 2.2: The angle η_k as defined for the k-th site.

2.2 Interplanetary Trajectory Optimization

The objective here is to design an interplanetary trajectory for a spacecraft to travel from the home planet to the target planet with a minimum cost. The spacecraft will benefit from as many as needed swing-bys of other planets. The spacecraft can also apply deep space maneuvers (DSMs). The segment between any two planets is called a leg. A leg can have any number of DSMs, see **Fig. 1.2**. The spacecraft can have multiple revolutions about a planet before it leaves to the next planet. The scenario of the mission refers to the sequence of swing-bys. The determination of the mission scenario then means the determination of the number of swing-bys, the planets of swing-bys, and the times of swing-bys. The problem is worked out in a two-body framework, and is formulated as an optimization problem.

The problem is formulated as follows: For given ranges for departure and arrival dates from the home planet to a target planet, find the optimal selections for the number of swing-bys, the planets to swing-by, the times of swing-bys, the number of DSMs, the amounts and directions of these DSMs, the times at which these DSMs are applied, and the exact launch and arrival dates, such that the total cost of the mission is a minimum.

Dependent Variables

An essential step in any genetic optimization algorithm is to evaluate the cost function at different design points. At each design point, the optimization algorithm selects values for all the independent design variables of the problem. There are several other dependent variables that need to be computed before the cost function can be evaluated. This section details how these dependent variables are computed, given a set of values for the independent design variables.

N-Impulses Maneuver

In the simple case of a Two-Impulse interplanetary orbit transfer, the total number of design variables is two. In this case, the trajectory has one leg between the home and arrival planets. The two variables are the departure and arrival dates. For a candidate solution the departure and arrival dates fix the time of flight T. A Lambert's problem is then solved to find the transfer orbit. The required departure and arrival impulses, $\Delta\vec{V}_d$ and $\Delta\vec{V}_a$, are then calculated.

In an N-impulses trajectory (no swing-bys), There are $n-1$ different orbits, and the number of DSMs is $n-2$, in addition to the departure and arrival impulses. The number of unknowns in this case is $4n$. These unknowns are the impulse velocity increment vectors $\Delta\vec{V}_i$ and the times of applying them (t_i), for all impulses, where $i = 1 : n$. The time of application of each DSM is defined as a fraction, ε, of the overall transfer time, T. So, $t_{DSM_j} = \varepsilon_j T$, where $0 < \varepsilon_j < 1$ and $j = 2 : n-1$. **Figure** 2.3 shows a three-impulses trajectory (one DSM).

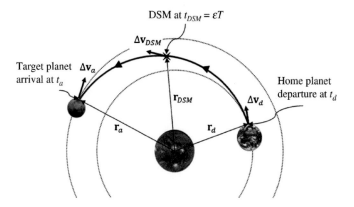

Figure 2.3: Three-impulses transfer orbit.

Of the $4n$ unknowns, there are $4n - 6$ independent design variables. These independent variables are selected to be: the departure and arrival dates, the velocity increment vector(s) at the first $n - 2$ impulses, and the fractional variable(s) ε_j at each DSM. The remaining 6 unknowns are computed as functions of the independent design variables as follows: the departure and arrival dates fix the home and target planets positions, and hence the spacecraft positions at these two locations, $\vec{r}_d$ and $\vec{r}_a$, are fixed. The velocity increment at departure, $\Delta\vec{V}_d$, yields the spacecraft initial velocity on the first transfer orbit, which along with $\vec{r}_d$ fixes the first transfer orbit. Once we have the first transfer orbit, and using the time of applying the fist DSM, ε_1, the location of the first DSM is computed using Kepler's equation [112]. The velocity of the spacecraft at that location before applying the DSM is also computed. The velocity increment at the first DSM is used to compute the spacecraft velocity right after the first DSM is applied. The procedure used in the first transfer orbit is repeated for all subsequent transfer orbits, but the last one. On the last transfer orbit, a Lambert's problem is solved. The spacecraft positions at the last DSM and at arrival are known. The time of flight is also known. A Lambert's problem solution yields the last transfer orbit. The velocity increments at the last DSM and at arrival are then computed. The total cost of the mission in this case is:

$$\Delta\vec{V} = \|\Delta\vec{V}_d\| + \sum_{i=1}^{k} \|\Delta\vec{V}_{DSM_i}\| + \|\Delta\vec{V}_a\| \tag{2.9}$$

where k is the number of deep space maneuvers.

Gravity Assist Maneuvers

In this analysis, a gravity assist maneuver is assumed to be an instantaneous impulse applied to the spacecraft. So, the spacecraft position vector is assumed not to change during the swing-by maneuver, and is equal to the swing-by planet position vector at the swing-by instance.

$$\vec{r}^- = \vec{r}^+ = \vec{r}_p \tag{2.10}$$

Where $\vec{r}^-$ and $\vec{r}^+$ are the spacecraft incoming and outgoing position vectors, respectively, and $\vec{r}_p$, is the swing-by planet position vector.

As shown in **Fig.** 2.4, the trajectory of the spacecraft relative to the planet is a hyperbola; the relative hyperbolic velocity $\vec{v}_\infty$ is defined as:

$$\vec{v}_\infty = \vec{v}_{s/c} - \vec{v}_p \tag{2.11}$$

where $\vec{v}_{s/c}$ and $\vec{v}_p$ are the spacecraft and planet velocity vectors at the swing-by instance, respectively.

Two types of swing-bys are implemented in this work: powered swing-bys and non-powered swing-bys [79]. In a non-powered swing-by, the incoming and outgoing relative velocities, $\vec{v}_\infty^-$ and $\vec{v}_\infty^+$, respectively, have the same magnitude.

$$\|\vec{v}_\infty^-\| = \|\vec{v}_\infty^+\| = v_\infty \tag{2.12}$$

The swing-by plane is defined by the incoming relative velocity vector and the pericenter radius vector $\vec{r}_{per}$. The change in the relative velocity direction in the swing-by plane, δ, can be computed from Eq. (2.13) [96].

$$\sin(\delta/2) = \frac{\mu_p}{\mu_p + r_{per}v_\infty^2} \tag{2.13}$$

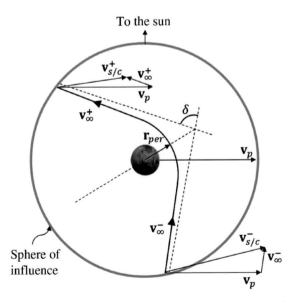

Figure 2.4: Gravity assist model as seen in the swing-by plane projection.

where μ_p is the gravitational constant of the swing-by planet, and $r_{per} = \|\vec{r}_{per}\|$. In a non-powered swing-by, the spacecraft velocity change is given by Eq. (2.14) [79].

$$\|\Delta\vec{v}_{nps}\| = \|\vec{v}_\infty^+ - \vec{v}_\infty^-\| = 2v_\infty \sin(\delta/2) \tag{2.14}$$

For a non-powered swing-by to be feasible, the periapsis radius must be higher than a minimum radius, i.e., $r_{per} > r_{min}$. In this work, we assumed $r_{min} = 1.1r_p$. The maximum deflection angle δ_{max} corresponds to $r_{per} = r_{min}$. If the desired gravity assist is not feasible via a non-powered swing-by, then an impulsive post-swing-by maneuver is applied (the powered swing-by). By applying a small impulse during the swing-by, higher deflection angles could be attained [90]. The powered swing-by impulse can be computed using Eq. (6.20).

$$\Delta\vec{v}_{ps} = (\vec{v}_{s/c}^+)_d - (\vec{v}_{s/c}^+)_{nps} \tag{2.15}$$

$(\vec{v}_{s/c}^+)_d$ is the desired spacecraft heliocentric outgoing velocity vector; and $(\vec{v}_{s/c}^+)_{nps} = \vec{v}_p - \vec{v}_\infty^+$.

The swing-by plane needs to be computed in order to calculate $\vec{v}_\infty^+$. To define the swing-by plane a rotation angle η is introduced [115]. The angle η represents the rotation of the ecliptic plane around the incoming relative velocity vector $\vec{v}_\infty^-$. The vector $\vec{v}_\infty^+$ is obtained form the vector $\vec{v}_\infty^-$ by performing two consecutive rotations (δ and η). The incoming relative velocity vector $\vec{v}_\infty^-$ is defined in the heliocentric inertial frame $\hat{I}\hat{J}\hat{K}$. A local frame $\hat{i}\hat{j}\hat{k}$ is defined such that the unit vector $\hat{i}$ is in the direction of the incoming relative velocity vector, so

$$\hat{i} = \frac{\vec{v}_\infty^-}{\|\vec{v}_\infty^-\|} \tag{2.16}$$

The plane of the swing-by maneuver is the $\hat{i}\hat{j}$ plane. Therefore, the outgoing relative velocity vector, as expressed in the local frame, $(\vec{v}_\infty^+)_L$, is:

$$(\vec{v}_\infty^+)_L = v_\infty[\cos\delta \quad \sin\delta \quad 0]^T \tag{2.17}$$

The $\vec{v}_\infty^+$ is computed in the inertial frame, via a coordinate transformation from the local frame to the inertial frame, as shown in **Fig.** 2.5. To perform this transformation, the following procedures are applied. The perpendicular plane Π defined by its normal $\hat{i}$ intersects with the inertial ecliptic (IJ plane) in the line Γ. The direction of Γ depends on the orientation of the incoming relative velocity in the inertial frame. The angle between Γ and the inertial $\hat{I}$ is Ω which is defined in the IJ plane. The angle between Γ and the unit vector $\hat{j}$ is the rotation angle η which is defined in the perpendicular plane Π. The inclination of plane Π to the ecliptic plane is i. A three angle rotation is performed to calculate the local unit vector $\hat{j}$ in the inertial frame according to the following relation:

$$\hat{j} = R_3(-\Omega)R_1(-i)R_3(-\eta)[1 \quad 0 \quad 0]^T \tag{2.18}$$

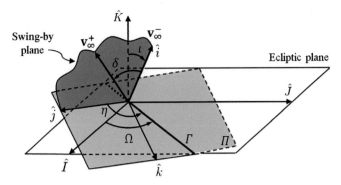

Figure 2.5: Transformation scheme from the local frame ijk to the inertial frame IJK showing the definition of the rotation angle η.

Finally, the outgoing velocity vector in the inertial frame is computed as follows:

$$\vec{v}_{\infty}^{+} = C(\vec{v}_{\infty}^{+})_L \tag{2.19}$$

where C is the transformation matrix which is calculated as in Eq. (2.20).

$$C = \begin{bmatrix} \hat{i} & \hat{j} & \hat{k} \end{bmatrix} \tag{2.20}$$

N-Impulses Multi Gravity Assist Maneuvers

In this section, a formulation is introduced for the full problem of optimal trajectory design, including multi gravity assist maneuvers, and possible N-impulses in each leg. The mission is designed to transfer a spacecraft from a home (departure) planet P_d to a target (arrival) planet P_a. Time windows are given for each of the departure and arrival dates. Consider a mission that consists of m gravity assist maneuvers. For the ith trajectory leg, there are n_i deep space maneuvers, where $i = 1 : m + 1$. The time of flight for each leg, except the last leg, is an independent design variable. The calculations of the dependent variables are carried out starting from the departure planet, and from one leg to the next and so on.

In any trajectory leg, the spacecraft trajectory is solved as discussed in **Section** 2.2. The velocity vector at the leg end point is the heliocentric incoming velocity vector of the consequent gravity assist maneuver. If a leg has at least one DSM, then the swing-by maneuver at the beginning of that leg is assumed a non powered swing-by maneuver. The swing-by independent design variables are the pericenter altitude h and the rotation angle η. The swing-by maneuver calculations are carried out, as discussed in **Section** 2.2, to yield the outgoing spacecraft velocity vector. This velocity vector is the spacecraft heliocentric velocity vector of the initial point on the consequent transfer trajectory. The process is repeated for all legs and swing-by maneuvers.

If any of the swing-by maneuvers is followed by a leg with no DSM, then that swing-by is assumed to be a powered swing-by maneuver. In this case, all dependent variables associated with the leg before the swing-by is calculated first. Then the dependent variables associated with the leg after the powered swing-by are computed. Finally the powered swing-by variables, including the swing-by impulse, are computed such that the two legs calculations are compatible. Assume the powered swing-by planet is planet $i + 1$. The leg before planet $i + 1$ is leg i, and the leg after it is leg $i + 1$. The times of swinging by planets $i + 1$ and $i + 2$ are known, and hence the positions of the spacecraft at these two times are known. There is no DSM in leg $i + 1$, hence a Lambert's problem is solved to calculate the spacecraft velocity vectors on the initial and final points of leg $i + 1$. The velocity vector at the initial point of leg $i + 1$ is the required heliocentric outgoing velocity vector $(\vec{v}_{s/c}^+)_d$ of the powered swing-by at planet $i + 1$. To achieve this velocity, a swing-by impulse $\Delta \vec{v}_{ps}$ is added. Since, it is desired to achieve all maneuvers with minimum fuel. We assume that the powered swing-by maneuver plane is the plane containing the incoming relative velocity vector $\vec{v}_\infty^-$ and the required outgoing relative velocity vector $(\vec{v}_\infty^+)_d$, so that $\Delta \vec{v}_{ps}$ does not have an out of plane component.

$$(\vec{v}_\infty^+)_d = (\vec{v}_{s/c}^+)_d - \vec{v}_p \qquad (2.21)$$

Equations (2.14, 6.20, and 2.21) are used to calculate $\Delta \vec{v}_{ps}$. The desired deflection angle δ_d between $\vec{v}_\infty^-$ and $(\vec{v}_\infty^+)_d$ is computed as:

$$\delta_d = \cos^{-1}\left(\frac{\vec{v}_\infty^- \cdot (\vec{v}_\infty^+)_d}{\|\vec{v}_\infty^-\| \|(\vec{v}_\infty^+)_d\|} \right) \qquad (2.22)$$

If $\delta_d \leq \delta_{max}$, then $\vec{v}_\infty^+$ is in the same direction as $(\vec{v}_\infty^+)_d$ and is calculated as:

$$\vec{v}_\infty^+ = \vec{v}_\infty \frac{(\vec{v}_\infty^+)_d}{\|(\vec{v}_\infty^+)_d\|} \qquad (2.23)$$

If $\delta_d > \delta_{max}$, then δ_{max} is used, and the swing-by maneuver is carried out in the same plane defined by $\vec{v}_\infty^-$ and $(\vec{v}_\infty^+)_d$.

The objective here is to minimize the total cost, Δv, of the trajectory for a MGADSM mission that consists of m gravity assist maneuvers and n deep space maneuvers. Equation (2.24) shows the total cost of the mission.

$$\Delta v = \|\Delta \vec{v}_d\| + \sum_{i=1}^{m} \|\Delta \vec{v}_{ps_i}\| + \sum_{j=1}^{n} \|\Delta \vec{v}_{DSM_j}\| + \|\Delta \vec{v}_a\| \qquad (2.24)$$

where $\Delta \vec{v}_d$ and $\Delta \vec{v}_a$ are the departure and arrival impulses, respectively, $\Delta \vec{v}_{ps}$ is the post-swing-by impulse of the powered gravity assist only, and $\Delta \vec{v}_{DSM}$ is the applied deep space maneuver impulse. The fitness F_i at a design point is defined as:

$$F_i = \frac{1}{\Delta v} \qquad (2.25)$$

Equation (2.25) represents the objective function that is to be optimized. The system architecture variables are the number of DSMs and the number of swing-bys. The number of DSMs and the number of swing-bys dictate the total number of variables in a given problem; hence this is an example of a VSDS optimization problem. It is also noted that some of the variables are integers, such as the planet of each swing-by, and some variables are continuous. There are several constraints in the problem; these constraints are mainly the boundaries on each variable.

2.3 Optimization of Wave Energy Converters

There are several concepts for harvesting the wave energy such as the heaving WEC, Oscillating Surge Wave Energy Converters (OSWEC) and overtopping devices, which are illustrated in **Fig.** 2.6.

A simple heaving WEC consists of a floating buoy connected to vertical hydraulic cylinders (spar) which are attached at the bottom to the seabed or to a large body whose vertical motion is negligible relative to the float. When the float moves due to waves, the hydraulic cylinders drive hydraulic motors which in turn drive a generator [77]. Forces on the float include the hydrostatic force (buoyancy and weight) that can be computed from the geometry of the float, the excitation forces (due to the wave field and the specific geometry of the buoy), and the radiation (added) forces due to the waves created by the motion of the float itself [22, 77]. Some of these quantities are frequency dependant, and the wave field has many frequencies. The widely used equation of motion for a simple 1-DoF (heave only) FSB WEC can be written as [29]:

$$m\ddot{z}(t) = \overbrace{\int_{-\infty}^{\infty} h_f(\tau)\eta(t-\tau,z)d\tau}^{\text{excitation force } f_e} + f_s - \mu\ddot{z}(t) - \overbrace{\int_{-\infty}^{t} h_r(\tau)\dot{z}(t-\tau)d\tau}^{\text{radiation force } f_r} - u \quad (2.26)$$

where z is the heave displacement of the buoy from the sea surface, t is the time, m is the buoy mass, u is the control force, and f_s is the difference between the gravity and buoyancy force and it reflects the spring-like effect of the fluid. The pressure effect around the immersed surface of the float is called the excitation force, f_e, where η is the wave surface elevation at buoy centroid and h_f is the impulse response function defining the excitation force in heave. The radiation force, f_r, is due to the radiated waves from the moving float, where μ is a frequency dependant added mass, and h_r is the impulse response function defining the radiation force in heave.

The radiation force f_r in Eq. (2.26) can be approximated using a state space model of N states, $\vec{x}_r = [x_{r1}, \cdots, x_{rN}]^T$, which outputs the radiation force [122]:

$$\dot{\vec{x}}_r = A_r\vec{x}_r + B_r\dot{z} \quad \text{and} \quad f_r = C_r\vec{x}_r, \quad (2.27)$$

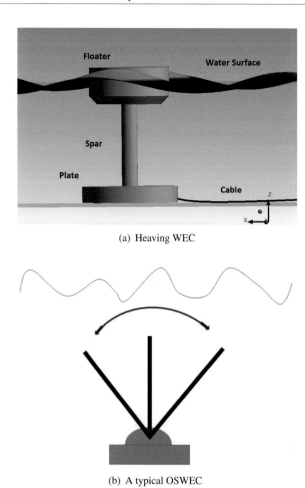

(a) Heaving WEC

(b) A typical OSWEC

Figure 2.6: Different types of wave energy converters.

where the constant radiation matrices A_r and B_r are obtained by approximating the impulse response function in the Laplace domain, as detailed in several references such as [93]. The impulse response function of the excitation force (h_f in Eq. (2.26)), however, is non-causal. This can be explained by considering that the output excitation force as well as the input incident wave elevation at the buoy origin have a distant primary cause (e.g., wave maker of a wave tank.) So, to compute the excitation force convolution integral in Eq. (2.26), future information is required on the wave elevation $\eta(t)$ (since $h_f(t) \neq 0$ for $t < 0$) [46]. This is the reason why most of the WEC control methods, developed over the past decades, require prediction in the future for the wave elevation (or the excitation forces), for a period t_H in the future that is sufficiently large to make $h_f(t)$ negligible for $t < -t_H$. In fact, the optimal control that maximizes the harvested energy, computed using the above model in the frequency domain, is known to be a function of the complex conjugate of the intrinsic mechanical impendence, which is a non-causal impulse response function [46].

To maximize the harvested energy, the optimal control problem is usually formulated as:

$$Min : J((z(t), u(t)) = \int_0^{t_f} \{-u(t)\dot{z}(t)\}dt, \text{ Subject to: } Equation \text{ (2.26)}. \quad (2.28)$$

Energy harvesting of a single WEC is all about motion control, in addition to shape optimization. Regarding the motion control, most of the existing literature present controls that are designed using a linear dynamic model, e.g., [47, 73, 72, 48, 100, 103]. Reference [77] implements dynamic programming while reference [64] uses a gradient based algorithm in searching for the optimal control. A model predictive control (MPC) can be used as in [28, 76, 88, 104, 75]. Reference [21] utilized the pseudo spectral method whereas references [11, 12] developed a shape-based approach that needs a fewer number of approximated states compared to the pseudo spectral method [42]. In the presence of limitations on the control actuation level, a bang-bang suboptimal control was proposed in [16]. There are few control approaches that do not require wave prediction such as the multi-resonant control approach [105, 13]. Most of the control approaches in the literature attempts to maximize the mechanical power of the WEC by searching for the optimal control force. Often, the computed control forces have a spring-like component.

Regarding the shape, there is not much done in the literature on the shape optimization, since any non-cylindrical shape terminates the validity of the linear model that is being used. There are multiple sources of possible nonlinearities in the WEC dynamic model [120]. For example, if the buoy shape does not have a constant cross section near the water surface then the hydrostatic force will be nonlinear. Other nonlinearities include: the coupling between the heave and pitch modes [117, 123], and nonlinearities due to large motion [60]. Reference [60] presents a numerical analysis for the nonlinear hydrodynamic forces at different levels from a full nonlinear model using computational fluid dynamics (CFD) tools, to linear models corrected by the nonlinear Froude Krylov force as well as nonlinear viscous and hydrostatic forces. The power take off (PTO) unit may have nonlinearities as well [20]. Reference [98] points out that different WEC systems should choose only the relevant nonlinear effects to avoid unnecessary computational costs. For example, in the case of heaving point absorbers the nonlinear Froude Krylov force is essential while the nonlinear diffraction and radiation can be neglected; the nonlinear viscous effects are weak as well for point absorbers [98]. The nonlinear PTO and mooring effects seem to be significant. In fact references [82, 99] focus on the nonlinear Froude Krylov forces and show that they are the dominant nonlinearities in the case of a heaving point absorber with nonuniform cross sectional area. Recently, reference [58] presented a computational way of evaluating the static and dynamic nonlinear Froude Krylov forces. Two methods are presented in [58]. The first method assumes that the wave length is considerably longer than the characteristic length of the device and hence ignores the dependance of the pressure (on the buoy surface) on the surge coordinate. The second method uses a McLaurin expansion to simplify the force integral calculation. The latter method was demonstrated to be more accurate for various sea states.

In this chapter, only the nonlinear Froude Krylov forces will be assumed for simplicity of the presentation. Only cylindrical buoys would generate linear hydrostatic forces. Other shapes add nonlinear forces to the model.

The buoy of a wave energy converter is usually axisymmetreic. Using an axisymmetric body design, it is possible to assume only one direction of the incoming exciting wave, and hence it is easier to analyze the system. It is also more convenient to compute analytical expressions for the nonlinear Froude Krylov forces on an axisymmetric buoy [59]. These nonlinear Froude Krylov forces depend on the shape, and hence the shape can be optimized to leverage these nonlinear forces in extracting more of the wave energy.

As shown in **Fig.** 2.7, the surface of an axisymmetric body can be described in parametric cylindrical coordinates $[\sigma, \theta]$ as generic profile $f(\sigma)$, where σ is the coordinate of a point with respect to z axis, θ is the angle oriented from the

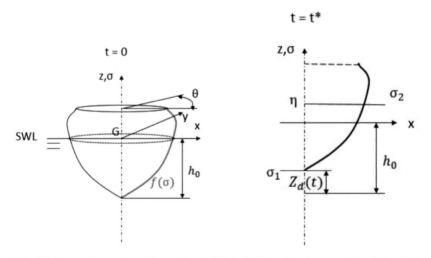

(a) Equilibrium position at the still water level (SWL) and the draft h_0

(b) Free elevation η and the device displacement z_d after a time t^*

Figure 2.7: The surface of an axisymmetric heaving device with generic profile $f(\sigma)$. The pressure is integrated over the wetted surface between σ_1 (the bottom point of the buoy) and σ_2 (the wave elevation at time t).

positive x axis direction to the position vector of a point [59]:

$$
\begin{aligned}
x(\sigma, \theta) &= f(\sigma)\cos\theta \\
y(\sigma, \theta) &= f(\sigma)\sin\theta \\
z(\sigma, \theta) &= \sigma \\
\theta &\in [0, 2\pi) \cap \sigma \in [\sigma_1, \sigma_2]
\end{aligned}
\tag{2.29}
$$

Using superposition, the total Froude Krylov force F_{FK} on a heaving axis, acting on a surface S, can be decomposed into smaller forces acting on corresponding areas, as follows:

$$
\begin{aligned}
F_{FK} &= \iint_S P\vec{n}\,dS \\
&= \int_0^{2\pi} \int_{\sigma_1}^{\sigma_2} Pf'(\sigma)f(\sigma)\,d\sigma\,d\theta \\
&= \int_0^{2\pi} \left[\sum_{i=1}^{N-1} \int_{\hat{\sigma}_i}^{\hat{\sigma}_{i+1}} Pf'(\sigma)f(\sigma)\,d\sigma \right] d\theta \\
&= \sum_{i=1}^{N-1} \int_0^{2\pi} \int_{\hat{\sigma}_i}^{\hat{\sigma}_{i+1}} Pf'(\sigma)f(\sigma)\,d\sigma\,d\theta
\end{aligned}
\tag{2.30}
$$

where $\hat{\sigma}_1 = \sigma_1, \hat{\sigma}_N = \sigma_2$, and P is the pressure on the wetted surface. A simplified expression of the non-linear Froude Krylov force is presented in [59] when the buoy shape is one of the four categories shown in **Fig.** 2.8. These expressions are as follows:

Vertical line

$$
f(\sigma) = R
\tag{2.31}
$$

$$
F_{FK_{st}} = \int_0^{2\pi} \int_{\sigma_1}^{\sigma_2} P_{st_z} f'(\sigma)f(\sigma)\,d\sigma\,d\theta = -\pi R^2 \rho g \sigma_1
\tag{2.32}
$$

$$
F_{FK_{dy}} = \int_0^{2\pi} \int_{\sigma_1}^{\sigma_2} P_{dy_z} f'(\sigma)f(\sigma)\,d\sigma\,d\theta = \pi R^2 \rho gae^{\chi\sigma_1}\cos\omega t
\tag{2.33}
$$

Oblique line

$$
f(\sigma) = m(\sigma - z_d) + q
\tag{2.34}
$$

$$
F_{FK_{st}} = \int_0^{2\pi} \int_{\sigma_1}^{\sigma_2} P_{st_z} f'(\sigma)f(\sigma)\,d\sigma\,d\theta = -2\pi\rho gm \left[m\frac{\sigma^3}{3} + (q - mz_d)\frac{\sigma^2}{2} \right]_{\sigma_1}^{\sigma_2}
\tag{2.35}
$$

$$
F_{FK_{dy}} = \int_0^{2\pi} \int_{\sigma_1}^{\sigma_2} P_{dy_z} f'(\sigma)f(\sigma)\,d\sigma\,d\theta = -\frac{2\pi}{\chi}\rho gam^2\cos\omega t \left[(\frac{q}{m} - \frac{1}{\chi} - z_d + \sigma)e^{\chi\sigma} \right]_{\sigma_1}^{\sigma_2}
\tag{2.36}
$$

Arc of circumference

$$f(\sigma) = \sqrt{R^2 - (\sigma - z_d^2)} \tag{2.37}$$

$$F_{FK_{st}} = \int_0^{2\pi} \int_{\sigma_1}^{\sigma_2} P_{st_z} f'(\sigma) f(\sigma) d\sigma d\theta = -2\pi\rho g \left[\frac{\sigma^3}{3} + z_d \frac{\sigma^2}{2} \right]_{\sigma_1}^{\sigma_2} \tag{2.38}$$

$$F_{FK_{dy}} = \int_0^{2\pi} \int_{\sigma_1}^{\sigma_2} P_{dy_z} f'(\sigma) f(\sigma) d\sigma d\theta = -\frac{2\pi}{\chi}\rho ga \cos \omega t \left[(z_d + \frac{1}{\chi} - \sigma) e^{\chi\sigma} \right]_{\sigma_1}^{\sigma_2} \tag{2.39}$$

Exponential profile

$$f(\sigma) = A e^{B(\sigma - z_d)} \tag{2.40}$$

$$F_{FK_{st}} = \int_0^{2\pi} \int_{\sigma_1}^{\sigma_2} P_{st_z} f'(\sigma) f(\sigma) d\sigma d\theta = -\pi\rho a A^2 \left[(\sigma - z_d - \frac{1}{2B}) e^{2B(\sigma - z_d)} \right]_{\sigma_1}^{\sigma_2} \tag{2.41}$$

$$F_{FK_{dy}} = \int_0^{2\pi} \int_{\sigma_1}^{\sigma_2} P_{dy_z} f'(\sigma) f(\sigma) d\sigma d\theta = 2\pi\rho g \frac{A^2 B}{2B + \chi} a e^{-2Bz_d} \cos \omega t \left[e^{(2B + \chi)\sigma} \right]_{\sigma_1}^{\sigma_2} \tag{2.42}$$

Using Eq. (2.30), however, the Froude Krylov force of a complex buoy shape can be computed using the simplified analytical equations in [59], if the complex buoy shape can be decomposed into sub-section, in which the shape of each sub-section is one of the four categories shown in **Fig.** 2.8. Decomposition of a complex shape generates several sections; each section S_i can be described by just two variables α_i and h_i, as shown in **Fig.** 2.8. The number of sections in the buoy shape is a variable; also the shape category of each section is a variable.

The control also can be parameterized. Since, this optimization approach excites a nonlinear phenomena, the following form of the control can be assumed:

$$u = \sum_{i=1}^{N_c} \alpha_{ci} z^i + \sum_{j=1}^{M_c} \beta_{cj} \dot{z}^j \tag{2.43}$$

where the coefficients α_{c_i}, β_{c_i}, α_{s_i}, and β_{s_i} are the design parameters, that are optimized for the given objective. The number of nonlinear stiffness terms N_c and the number of nonlinear damping terms M_c are both design parameters.

The optimization problem can then be formulated as follows. Find the extremal of the objective function defined in (2.28), subject to the constraints given in Eq. (2.26). The variables are the control coefficients α_{c_i}, β_{c_i}, α_{s_i}, β_{s_i}, the number of shape sections, and the shape category in each section. This is a VSDS problem.

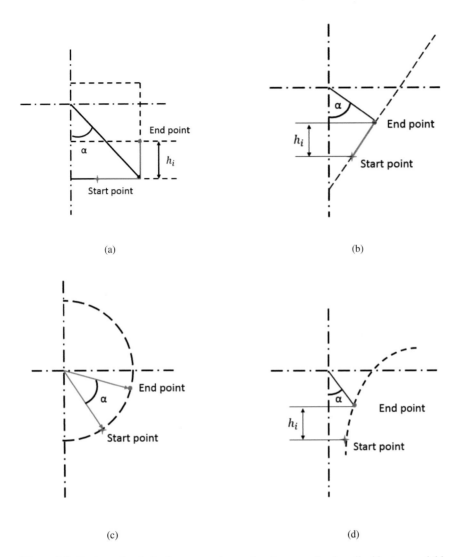

Figure 2.8: Each section i of a decomposed complex shape can be described by two variables α_i and h_i or less.

CLASSICAL OPTIMIZATION ALGORITHMS

Chapter 3

Fundamentals and Core Algorithms

Part I presented the necessary conditions of optimality. It was also shown that for relatively simple optimization problems, the necessary conditions for optimality can be used to find the optimal solution. For most practical optimization problems, numerical methods are needed to search for the optimal solution. This chapter presents basic numerical methods that are designed to solve relatively simple problems numerically. These algorithms however are building blocks that are usually used in more complex numerical algorithms that will be presented in subsequent chapters in **Part** II.

Consider the case of an unconstrained optimization problem that has only one optimization variable x, and the objective function is $f(x)$. The necessary condition for optimality in this case is:

$$\phi(x) \equiv \frac{df}{dx} = 0 \tag{3.1}$$

Solving the necessary condition for optimality can be thought of as a root finding problem to find the roots of $\phi(x) = 0$. Numerical algorithms for root finding include the Newton-Raphson and the bisection algorithms. The Newton-Raphson algorithm starts with an initial guess x_0 and updates this guess over subsequent iterations using the gradient $d\phi(x)/dx$. The bisection algorithm does not need the gradient information. It starts with two points x_1 and x_2 selected such that $\phi(x_1)\phi(x_2) < 0$ (hence x_1 and x_2 are on two opposite sides of the root), and repeatedly halves the interval between them with a new point x_3. In each iteration, the new point x_3 is used to update one of the interval boundaries x_1 or x_2 depending on the sign of $\phi(x_3)$. The focus in this chapter, however, is not on the root finding algorithms. Rather the numerical algorithms presented in this chapter searches for the maximum of $f(x)$.

Suppose the objective function $f(x)$ to be maximized is a function of only one variable x. Assume also that the there is only one maximum in the search domain $[x_{min}, x_{max}]$ (that is $f(x)$ is a unimodal function). This chapter presents different algorithms that numerically search for the maximum point of $f(x)$.

3.1 Equal Interval Search Algorithm

The first algorithm presented in this chapter and is probably the simplest one of them, is the qual Interval Search Algorithm. Given the boundaries of the search space, x_{min} and x_{max}, two intermediate points are computed as follows:

$$x_1 = \frac{x_{min} + x_{max}}{2} + \sigma$$

$$x_2 = \frac{x_{min} + x_{max}}{2} - \sigma$$

(3.2)

where the interval σ is defined as shown in **Fig.** 3.1.

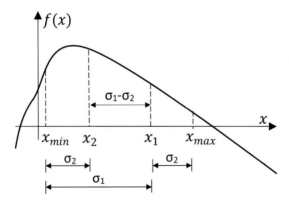

Figure 3.1: The Equal Interval Search Algorithm uses two intermediate points x_1 and x_2 to updtate one of the boundary points of the search space, at each iteration.

The function values at the two intermediate points, $f(x_1)$ and $f(x_2)$, are then evaluated. The search domain is then updated as shown in **Algorithm** 3.1.

The efficiency of the Equal Interval Search Algorithm depends on the selection of the interval σ. For a small value of σ, the algorithm might be inefficient.

Example 3.1. Consider the trapezoid cross section shown in **Fig.** 3.2. Each side is of unit length, and the base is also of unit length. Find the angle θ that maximizes the area of the trapezoid cross section.

Algorithm 3.1 Equal Interval Search Algorithm

Select an initial guess for the boundary points x_{min} and x_{max}
Select the interval σ
while a termination condition is not satisfied **do**
  Compute intermediate points x_1 and x_2 using Eq. (3.2)
  Evaluate the objective function at the intermediate points $f(x_1)$ and $f(x_2)$
  if $f(x_2) > f(x_1)$ **then**
    $x_{max} = x_1$
  end if
  if $f(x_2) < f(x_1)$ **then**
    $x_{min} = x_2$
  end if
  Update the search domain with the new interval $[x_{min}, x_{max}]$
end while

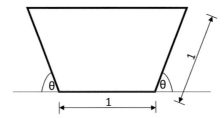

Figure 3.2: Trapezoid cross section.

Solution: The area of this trapezoid can be computed as function of the angle θ as:

$$A(\theta) = \sin\theta\,(1+\cos\theta) \tag{3.3}$$

To use the Equal Interval Search Algorithm, we need to select an initial range for the angle θ. The largest possible range is $[0, \pi/2]$ and is here selected as the initial guess; hence $x_{min} = 0$ and $x_{max} = \pi/2$. The interval σ is selected to be 0.2. The stopping criterion selected to terminate the algorithm is when the difference between x_{min} and x_{max} is sufficiently small. Samples of the numerical results of subsequent iterations are listed in **Table** 3.1. As can be seem from **Table** 3.1, after ten iterations the difference between x_{min} and x_{max} is about 0.2 radians which is considered large. To increase the speed of convergence of the method we change the value of the interval σ to be 0.02 and continue. At iteration number twenty the difference between x_{min} and x_{max} reduces to about 0.02 radians. For more accurate solution, we reduce the interval to 0.005 and continue. At iteration number thirty, the difference between x_{min} and x_{max} is about 0.005 radians. If further improvement is needed, we can reduce the interval more

Table 3.1: Iterations of the equal interval search algorithm for **Example** 3.1.

Iteration	x_1	x_2	$f(x_1)$	$f(x_2)$	x_{min}	x_{max}
1	0.885398163	0.685398163	1.264200367	1.123014596	0.685398163	1.570796327
2	1.228097245	1.028097245	1.258341777	1.29856249	0.685398163	1.228097245
10	1.149115816	0.949115816	1.285843539	1.286336023	0.947777148	1.149115816
11	1.058446482	1.038446482	1.298874088	1.298938458	0.947777148	1.058446482
20	1.057206862	1.037206862	1.298908214	1.298908198	1.037206862	1.057383951
21	1.049795406	1.044795406	1.299029343	1.299030606	1.037206862	1.049795406
30	1.049691657	1.044691657	1.299030029	1.299029944	1.044691657	1.049706478
31	1.047449067	1.046949067	1.299038024	1.299038025	1.044691657	1.047449067
40	1.047444658	1.046944658	1.299038026	1.299038023	1.046944658	1.047449067

to 0.0005 and run for more iterations, which yields a difference of 0.0005 radians in iteration forty. This corresponds to a solution of trapezoid area of 1.299 unit area, and an angle $\theta = 1.047$ radians. **Figure** 3.3 shows the convergence history in this problem as we change the interval value. This concept of changing an optimization parameter in subsequent iterations will be further utilized in the following chapters in more complex situations.

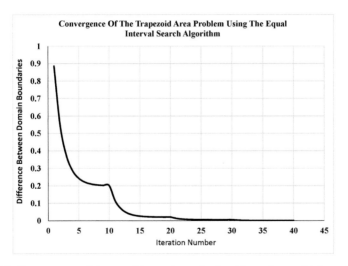

Figure 3.3: Convergence of the equal interval search algorithm using multiple vaules for the interval σ.

3.2   Golden Section Method

The Golden Section Algorithm is more efficient than the Equal Interval Search Algorithm. The key difference between the Golden Section Algorithm and the Equal Interval Search Algorithm is in the criteria for selecting the intermediate points. Consider the problem of maximizing the function $f(x)$ in the range $x_{min} < x < x_{max}$. In the Golden Section Algorithm the first intermediate point x_1 is selected such that:

$$\frac{\sigma_1}{\sigma_1 + \sigma_2} = \frac{\sigma_2}{\sigma_1} \qquad (3.4)$$

where σ_1 and σ_2 are defined in **Fig.** 3.4.

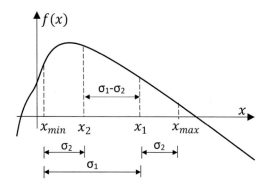

Figure 3.4: The Golden Section Algorithm selects the two intermediate points x_1 and x_2 using the Golden Ratio criterion.

The second intermediate point is selected in a similar way in the larger of $[x_{min}, x_1]$ and $[x_1, x_{max}]$ such that:

$$\frac{\sigma_2}{\sigma_1} = \frac{\sigma_1 - \sigma_2}{\sigma_2} \qquad (3.5)$$

Note that the rations in Eqs. (3.4) and (3.5) are equal and have a special value known as the Golden Ratio. Let R be the ration $R = \sigma_2/\sigma_1$, then Eq. (3.4) can be rewritten as:

$$\frac{\sigma_1 + \sigma_2}{\sigma_1} = \frac{\sigma_1}{\sigma_2}$$

$$\therefore 1 + R = \frac{1}{R}$$

$$\therefore R^2 + R - 1 = 0$$

$$\therefore R = \frac{\sqrt{5} - 1}{2} = 0.61803 \qquad (3.6)$$

The Golden Ratio is $1 + R = 1.61803$. Note that since R is a constant ratio and $\sigma_1 + \sigma_2 = x_{max} - x_{min}$, then we can rewrite Eq. (3.4) as follows:

$$\sigma_1 = R(\sigma_1 + \sigma_2) = R(x_{max} - x_{min}) \qquad (3.7)$$

Hence we can write:

$$x_1 = x_{min} + \sigma_1 = x_{min} + R(x_{max} - x_{min}) = (1 - R)x_{min} + Rx_{max} \qquad (3.8)$$

In a similar way we can write:

$$x_2 = Rx_{min} + (1 - R)x_{max} \tag{3.9}$$

Once the two intermediate points are computed, the function is evaluated at the two intermediate points. One of the boundary points is then updated depending on the values of $f(x_1)$ and $f(x_2)$ as follows. If $f(x_2) > f(x_1)$ then the new x_{max} in the next iteration is set to the current x_1, that is $x_{max}^i = x_1^{i-1}$ where the i denotes the iteration number. In this case also the new x_{min} remains unchanged and the new first intermediate point is selected to be the previous second intermediate point, that is $x_{min}^i = x_{min}^{i-1}$ and $x_1^i = x_2^{i-1}$. If $f(x_2) < f(x_1)$ then the new x_{min} in the next iteration is set to the current x_2, that is $x_{min}^i = x_2^{i-1}$. In this case also the new x_{max} remains unchanged and the new second intermediate point is selected to be the previous first intermediate point, that is $x_{max}^i = x_{max}^{i-1}$ and $x_2^i = x_1^{i-1}$. With the new three points selected (x_{min}, x_1 and x_{max}) the fourth point is selected as described above to satisfy the Golden Ration criterion. This process repeats until a stopping criterion is satisfied. One stopping criterion could be that the distance between the boundary points $\Delta x = x_{max} - x_{min}$ is sufficiently small. Another stopping criterion could be a maximum number of iterations. A rigorous algorithm would have both criterion implemented where the algorithm stops once one of the two criteria is satisfied. **Algorithm** 3.2 is the Golden Section Algorithm.

Algorithm 3.2 Golden Section Algorithm

Select an initial guess for the boundary points x_{min} and x_{max}
Select a stopping interval Δx and a maximum number of iterations N_{itr}
Compute intermediate points x_1 and x_2 using Eqs. (3.8) and (3.9)
i = 1
while $x_{max} - x_{min} > \Delta x$ and $i < N_{itr}$ **do**
    Evaluate the objective function at the intermediate points $f(x_1)$ and $f(x_2)$
    if $f(x_2) > f(x_1)$ **then**
        $x_{max} = x_1$
        $x_1 = x_2$
        Compute x_2 using Eq. (3.8)
    end if
    if $f(x_2) < f(x_1)$ **then**
        $x_{min} = x_2$
        $x_2 = x_1$
        Compute x_1 using Eq. (3.9)
    end if
    i = i + 1
end while

Example 3.2. Solve the problem in **Example** 3.1 using the Golden Section Algorithm.

Solution: The numerical results for sample iterations are listed in **Table** 3.2. Also **Fig.** 3.5 shows the convergence of the algorithm. Comparing the Golden Section Algorithm to the Equal Interval Search Algorithm we can see that the convergence is significantly faster using the Golden Section Algorithm, in addition to eliminating the need to update any parameters during the iterative process. Specifically, the interval size is reduced to below 0.0005 in only seventeen iterations using the Golden Section Algorithm compared to forty iteration when using the Equal Interval Search Algorithm.

Table 3.2: Iterations of the Golden Section Algorithm for **Example** 3.2.

Iteration	x_1	x_2	$f(x_1)$	$f(x_2)$	x_{min}	x_{max}
1	0.970805519	0.599990807	1.291357017	1.030651098	0.599990807	1.570796327
2	1.199981615	0.970805519	1.269777571	1.291357017	0.599990807	1.199981615
9	1.045571452	1.037678203	1.29903467	1.298920176	1.037678203	1.058342998
16	1.047434796	1.047162938	1.299038033	1.299038104	1.046723062	1.047434796
17	1.047162938	1.04699492	1.299038104	1.299038052	1.04699492	1.047434796

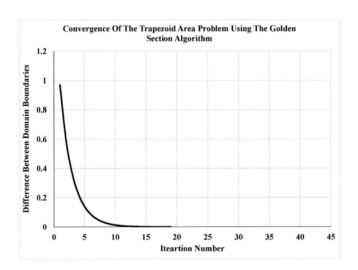

Figure 3.5: Convergence of the Golden Section Algorithm.

3.3 Linear versus Nonlinear Optimization

A function $h(\vec{x})$ is said to be linear if it satisfies Eq. (3.10).

$$h(a\vec{x}+b\vec{y}) = ah(\vec{x})+bh(\vec{y}) \tag{3.10}$$

where a and b are scalars, the vectors $\vec{x}$ and $\vec{y}$ are two arbitrary points in the space $\mathbb{R}^n$ (in the case of design optimization that aims to minimize f, each of $\vec{x}$ and $\vec{y}$ represents a design in the design space).

An optimization problem is said to be linear if all the functions in (1.1) are linear, and hence it takes the form:

$$\begin{aligned}&\text{minimize: } \vec{c}^T\vec{x}\\&\text{subject to: } \vec{a}_i^T\vec{x} \le b_i, i = 1\cdots m.\end{aligned} \tag{3.11}$$

where each of the vectors $\vec{c}$ and $\vec{a}_i, \forall i$, is a vector of scalars. Note that a linear equality constraint can always be mathematically eliminated from the problem formulation by eliminating one of the design variables. For example, if a linear constraint is given as $x+y = 0$, then it is possible to substitute for $x = -y$ in all the functions, and hence eliminate x and the constraint from the mathematical formulation. If one or more of the functions $f(\vec{x})$, $g_i(\vec{x})$, and $h_j(\vec{x})$ in (1.1) is non-linear, then the problem is a nonlinear optimization problem. One special case of the nonlinear programming problems is the least-squares optimization problem. In least-squares problems, there are no constraint functions. The objective function is of the form:

$$\text{minimize: } f(\vec{x}) = \sum_{i=1}^{n}\left(\vec{a}_i^T\vec{x} - b_i\right)^2 \tag{3.12}$$

This least-squares problem is found in numerous applications, such as curve fitting problems. In fact, the least-squares is a very old subject. Gauss published on least-squares in 1820s, and his original work was translated from Latin to English by G.W. Stewart in 1995 [51, 52]. Least-squares problems have an analytic solution. The objective function in (3.12) can be expressed in a matrix form $f(\vec{x}) = ([A]\vec{x} - \vec{b})^2$, where a row i in the matrix $[A]$ includes the elements of $\vec{a}_i$ and the vector $\vec{b}$ includes the scalars b_i. The analytic solution of the above least-squares problem takes the form:

$$\vec{x} = \left([A]^T[A]\right)^{-1}[A]^T\vec{b} \tag{3.13}$$

Another special type of nonlinear optimization is the convex optimization problems. Consider the linearity condition (3.10). A function $h(\vec{z})$ is said to be convex if:

$$h(a\vec{x}+b\vec{y}) \le ah(\vec{x})+bh(\vec{y}) \tag{3.14}$$

where $a+b=1$, $a \geq 0$, and $b \geq 0$. The optimization problem (1.1) is said to be convex if the functions in (1.1) are convex. Unlike the least-squares problem, the convex optimization problem does not have an analytic solution. Yet there are methods that can solve this problem, such as the Interior-point methods. Reference [26] presents in details convex optimization problems and the methods to solve this type of problem.

The general nonlinear optimization problem is challenging. There is no known method that will find the optimal solution to the general problem. Numerical algorithms that search for the optimal solution of a problem can be categorized into local search methods such as gradient-based algorithms [26] and global search methods such as genetic algorithms (GAs) [62], particle swarm optimization [69], ant colony optimization [33], and differential evolution [94]. The applications in system architecture optimization discussed in this book are usually nonlinear optimization problems. Many practical design optimization problems are usually replete with local minima and hence a global search algorithm is usually needed for optimizing the system variables.

3.3.1 Linear Programming

Linear programming is an important class of problems which mathematical models are made up of functions that are linear. There is a wide range of applications in which the engineering optimization problem can be formulated as a linear programming problem. Some of the concepts used in solving linear programming problems are also used in deriving the necessary conditions of optimality for nonlinear programming problems; this section presents a brief overview for the Simplex method for linear programming problems. A more detailed presentation can be found in several references such as [116].

Consider a linear programming problem in which we aim to maximize a linear function $f(x,y)$ subject to the boundary conditions $x \geq 0$, $y \geq 0$, and the two inequality constraints:

$$
\begin{aligned}
g_1(x,y): \quad & y \leq -0.5x+4 \\
g_2(x,y): \quad & y \geq 0.5x+1
\end{aligned}
\tag{3.15}
$$

Figure 3.6 shows the design space for the x and y variables. The contour plot of the objective function is represented by straight lines in the design space. The lines representing $f(x,y)$ are plotted in black, each line corresponds to all points in the design space that has the same value of f. The arrow points to the direction of increasing $f(x,y)$. The constraint function g_1 is represented by a region in the design space that is bounded by the line $y = -0.5x+4$, as shown in **Fig.** 3.6. The constraint function g_2 is represented by a region in the design space that is bounded by the line $y = 0.5x+1$. The intersections of these lines and the boundaries determine the feasible region, and a set of intersection points. Since the objective function f is linear, then the maximum value of f will be at one

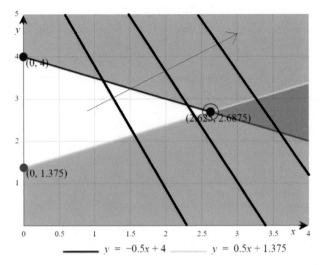

Figure 3.6: All functions are represented by lines in the design space in linear programming.

of these intersection points. Hence, one solution method for linear programming problems is to evaluate $f(x,y)$ at the intersection points and find the point of maximum $f(x,y)$. In doing that, one might evaluate f at one intersection point, and then check if there is a feasible change in any of the variables that will result in a higher value of f. This is the main concept of the Simplex Method which is widely used to solve this type of problem.

3.3.2 Nonlinear Programming

Nonlinear programming defines a collection of numerical methods for optimization problems whose mathematical models are nonlinear; that is when one or more of the functions $f(\vec{x})$, $g_i(\vec{x})$, and $h_j(\vec{x})$ in (1.1) is nonlinear. One type of particular importance in nonlinear programming problems is the Linear Quadratic Problems in which the objective function $f(\vec{x})$ is quadratic, while the constraint functions $g_i(\vec{x})$, and $h_j(\vec{x})$ are linear, $\forall i, j$. The solution of a nonlinear programming problem could be on the boundary of the feasible domain, or in the interior of the feasible domain in which case the solution has to satisfy the *necessary* conditions of optimality. An optimal solution in the interior of the feasible domain also satisfies the *sufficient* conditions of optimality for the given problem. For unconstrained nonlinear programming problems, the necessary conditions of optimality can be written as:

$$\vec{\nabla} f(\vec{x}^*) = \vec{0} \tag{3.16}$$

where $\vec{x}^*$ denotes the optimal solution, and $\vec{0}$ is a vector in which all components take the value of zero. The condition (3.16) is a necessary condition for optimal solutions in the interior of the feasible region. If the optimal solution is on the boundary, then it does not have to satisfy (3.16). The condition (3.16) states that the gradient of the objective function at the optimal solution should vanish. The condition (3.16) is sometimes referred to as the First Order Condition (FOC). For unconstrained nonlinear programming problems, a sufficient condition for optimality is satisfied if:

The Hessian matrix at the optimal solution, $[H(\vec{x}^*)]$, is positive definite. (3.17)

That is the eigenvalues of $[H(\vec{x}^*)]$ are all positive.

For constrained nonlinear programming problems, each type of the equality and inequality constraints is handled differently. Consider a nonlinear programming problem with only equality constraints, $h_i(\vec{x}), i = 1 \cdots l$. This problem can be transformed to an unconstrained optimization problem via the method of Lagrange as follows. A new function, called The Lagrangian $F(\vec{x}, \vec{\lambda})$, is introduced:

$$
\begin{aligned}
F(\vec{x}, \vec{\lambda}) &= f(\vec{x}) + \vec{\lambda}^T \vec{h}(\vec{x}) \\
&= f(\vec{x}) + \sum_{i=1}^{l} \lambda_i h_i(\vec{x})
\end{aligned}
\tag{3.18}
$$

The optimization problem becomes:

$$
\begin{aligned}
\text{minimize:} \quad & F(\vec{x}, \vec{\lambda}) \\
\text{subject to the boundaries:} \quad & \vec{x}^l \leq \vec{x} \leq \vec{x}^u
\end{aligned}
\tag{3.19}
$$

Despite that the new optimization problem in (3.19) is uncosntrained, the number of design variables is now $l + n$ (the variables are $\vec{\lambda}$ and $\vec{x}$). An optimal solution ($\vec{x}^*$ and $\vec{\lambda}^*$) in the interior of the feasible region must satisfy the necessary conditions (3.16) and the sufficient condition (3.17), for $F(\vec{x}, \vec{\lambda})$.

If the nonlinear programming problem has inequality constraints $g_i(\vec{x})$, one might consider transforming these inequality constraints to equality constraints by adding slack variables, as was the case in the Simplex method in **Section** 3.3.1. In this case, however, we add the squares of the slack variables since the latter are not restricted to be positive. So, we define new equality constraints functions

$$
\tilde{h}_i(\vec{x}, z_i) = g_i(\vec{x}) + z_i^2, \quad i = 1, \cdots, m.
\tag{3.20}
$$

The new unconstrained optimization problem is to minimize the Lagrangian $F(\vec{x}, \vec{\beta}, \vec{z})$:

$$
\text{minimize:} \quad F(\vec{x}, \vec{\beta}, \vec{z}) = f(\vec{x}) + \sum_{i=1}^{m} \beta_i \tilde{h}_i(\vec{x}, z_i)
\tag{3.21}
$$

$$
\text{subject to the boundaries:} \quad \vec{x}^l \leq \vec{x} \leq \vec{x}^u
$$

We can write the necessary conditions for optimality for the optimization problem (3.21) as follows:

$$\frac{\partial F}{\partial x_i} \equiv \frac{\partial f}{\partial x_i} + \sum_{j=1}^{m} \beta_j \frac{\partial g_j}{\partial x_i} = 0, \quad \forall i = 1 \cdots n$$

$$\frac{\partial F}{\partial z_i} \equiv 2\beta_i z_i = 0, \quad \forall i = 1 \cdots m \qquad (3.22)$$

$$\frac{\partial F}{\partial \beta_i} \equiv g_i + z_i^2 = 0, \quad \forall i = 1 \cdots m$$

where, $x_i = \vec{x}(i)$, and we have substituted for $\frac{\partial \bar{h}_j}{\partial x_i} = \frac{\partial g_j}{\partial x_i}, \forall i = 1 \cdots m$. The last two conditions in Eq. (3.22) can be combined as follows:

$$2\beta_i z_i = 0 \Rightarrow \quad \beta_i z_i^2 = 0, \quad \forall i = 1 \cdots m. \qquad (3.23)$$

$$g_i + z_i^2 = 0 \Rightarrow \quad g_i = -z_i^2 \Rightarrow \quad \beta_i g_i = -\beta_i z_i^2, \quad \forall i = 1 \cdots m. \qquad (3.24)$$

Combining Eq. (3.23) and Eq. (3.24) we get:

$$\beta_i g_i = 0, \quad \forall i = 1 \cdots m. \qquad (3.25)$$

The conditions in Eq. (3.25) are the necessary conditions for optimality associated with the inequality constraints in nonlinear programming problems. For the general nonlinear programming problem in (1.1), in which there are both equality and inequality constraints, the problem is transformed to an unconstrained optimization problem, in which the Lagrangian to be minimized is defined as:

$$F(\vec{x}, \vec{\beta}, \vec{\lambda}) = f(\vec{x}) + \vec{\lambda}^T \vec{h}(\vec{x}) + \vec{\beta}^T \vec{g}(\vec{x}) \qquad (3.26)$$

where $\vec{\lambda} = [\lambda_1, \cdots, \lambda_l]^T$, $\vec{\beta} = [\beta_1, \cdots, \beta_m]^T$, $\vec{h} = [h_1, \cdots, h_l]^T$, and $\vec{g} = [g_1, \cdots, g_m]^T$. The total number of variables is $(n + l + m)$, and the first order conditions for optimality of this general form optimization problem are:

$$\frac{\partial f(\vec{x}^*)}{\partial x_i} + \sum_{k=1}^{l} \lambda_k^* \frac{\partial h_k(\vec{x}^*)}{\partial x_i} + \sum_{j=1}^{m} \beta_j^* \frac{\partial g_j(\vec{x}^*)}{\partial x_i} = 0, \quad \forall i = 1 \cdots n$$

$$h_i(\vec{x}^*) = 0, \quad \forall i = 1 \cdots l \qquad (3.27)$$

$$\beta_i g_i(\vec{x}^*) = 0, \quad \forall i = 1 \cdots m$$

The conditions in (3.27) are called the KarushKuhnTucker (KKT), also known as the Kuhn-Tucker Conditions in honor of Harold W. Kuhn (1925-2014) and Albert W. Tucker (1905–1995) who formulated them in 1951. It was discovered later that these conditions had been stated by William Karush in his master's thesis in 1939. The optimal solution must satisfy the KKT conditions if it is in the interior of the feasible region; if it is on the boundary of the feasible region then it does not have to satisfy the KKT conditions.

Given m inequality constraints, let the index set of active constraints be $J_{ac}(\vec{x}^*) := \{j \in \{1, \cdots, m\} | g_j(\vec{x}^*) = 0\}$. Its compliment set (the set of inactive constraints) is $J_{in}(\vec{x}^*) := \{1, \cdots, m\} \backslash J_{ac}(\vec{x}^*)$. The last condition in (3.27) implies that either $\beta_i = 0$ or $g_i(\vec{x}^*) = 0$, $\forall i = 1, \cdots, m$. It can be shown that in the latter case (that is when $i \in J_{ac}(\vec{x}^*)$,) $\beta_i > 0$. The latter is called the strict complimentary slackness at $\vec{x}^*$. Let the size of $J_{ac}(\vec{x})$ be S_J (i.e., $S_J := |J_{ac}(\vec{x})|$,) and let $J(\vec{x}) = \{j_1, j_2, \cdots, j_{S_J}\}$, then we define the matrix $G(\vec{x})$ as:

$$G(\vec{x}) := \left[\vec{\nabla} h_1(\vec{x}), \vec{\nabla} h_2(\vec{x}), \cdots, \vec{\nabla} h_l(\vec{x}), \vec{\nabla} g_{j_1}(\vec{x}), \vec{\nabla} g_{j_2}(\vec{x}), \cdots, \vec{\nabla} g_{j_{S_J}}(\vec{x}) \right] \quad (3.28)$$

The second order sufficient optimality conditions of the optimization problem (1.1) are presented in several references, and can be written as follows. At the optimal solution $\vec{x}^*$:

1. Rank of $G(\vec{x}^*)$: the columns of $G(\vec{x}^*)$ are linearly independent

2. Strict Complementary Slackness: Strict Complementary Slackness holds at $\vec{x}^*$

3. The Hessian matrix: the Hessian of F with respect to $\vec{x}$ at $\vec{x}^*$ is positive definite; i.e., $\vec{d}^T [H] \vec{d} > 0, \forall \vec{d} \neq 0$, and $G(\vec{x}^*)^T \vec{d} = \vec{0}$

The numerical techniques that actually search for the optimal solution are not yet discussed.

Chapter 4

Unconstrained Optimization

This chapter presents some numerical algorithms that can be used to solved unconstrained optimization problems. The algorithms used to solve this type of problem can be categorized into two main categories: the gradient-based algorithms and the non gradient-based algorithms. Algorithms in the former class of algorithms need to compute gradient of the objective function with respect to each of the design variables. In the latter class of algorithms, on the other hand, gradient information is not needed. While gradient information helps the convergence to the optimal solution, it is not always easy to compute. Both categories of method are being widely used in different applications. The nonlinear optimization problem considered in this chapter can be written as:

$$\text{Minimize: } f(\vec{x}) \tag{4.1}$$

where $\vec{x} \in \mathbb{R}^n$, $\mathbb{R}^n$ is the n-dimensional Euclidian space, and $f : \mathbb{R}^n \longrightarrow \mathbb{R}$.

4.1 Non-Gradient Algorithms

Algorithms in this category of methods start with an initial guess for the solution $\vec{x}^0$, and updates that guess iteratively until a stopping criteria is satisfied. The iterative update equation takes the form:

$$\vec{x}^{k+1} = \vec{x}^k + \alpha^k \vec{s}^k \tag{4.2}$$

where, $\vec{x}^k$ is the current guess for the solution $\vec{x}^{k+1}$ is the next guess for the solution, $\vec{s}^k$ is the search direction at the current iteration, α^k is a scalar to be optimized in each iteration and it determines the step size. Different methods have different strategies in selecting the search direction in each iteration, but most of them use the same approach in computing the step size α^k at each iteration. Suppose that the search direction $\vec{s}^k$ at the kth iteration is given, then the new iterate can be computed using Eq. (4.3) and it will be a function of the step size $\vec{x}^{k+1}(\alpha^k)$. The objective function f is then evaluated at the new iterate and hence the objective function becomes a function of the step size $f(\alpha^k)$. The function $f(\alpha^k)$ is a function of only one variable α^k, and it can be optimized over α^k to find the optimal step size. Optimizing a function of one variable can be carried out either analytically or numerically using several algorithms such as the Golden Section method. This process for computing the step size is repeated in each iteration. Below is a description for some of the basic non-gradient algorithms.

4.1.1 Cyclic Coordinated Descent Method

In this method, the iterations are organized in cycles. For $\vec{x} \in \mathbb{R}^n$, each cycle will have n iterations. In each iteration in the cycle, the search direction $\vec{s}$ is selected to be unit vector of one of the design variables. In subsequent iterations in a cycle, the search direction cycles through the number of variables in order. The search directions repeat in subsequent cycles. For example, if we have three design variables, $n = 3$. Then each cycle will have three iterations. In each cycle, the first search direction $\vec{s} = [1,0,0]^T$, the second search direction is $\vec{s} = [0,1,0]^T$, and the third search direction is $\vec{s} = [0,0,1]^T$.

4.1.2 Pattern Search Method

In some problems, the Cyclic Coordinated Descent Method might get trapped in a zigzag pattern when the iterate gets close to the optimal solution. To overcome this problem, the Pattern Search Method adds one more iteration to each cycle. This additional iteration at the end of each cycle is computed as the summation of all the previous n search directions weighted by their step sizes; that is:

$$\vec{s}^{n+1} = \sum_{i=1}^{n} \alpha^i \vec{s}^i \tag{4.3}$$

This added iteration in each cycle disturbs the zigzag pattern.

Example 4.1. Consider the cantilever bean shown in **Fig.** 4.1. Due to the tip load Q, the tip point will be displaced and rotated. It is required to compute the linear displacement x_1 and the rotational displacement θ_2.

The displacement of a beam can be analyzed assuming that each end of the beam has two degrees of freedom: the vertical displacement and the rotational displacement, as shown in **Fig.** 4.1. In the case of a clamped cantilever beam,

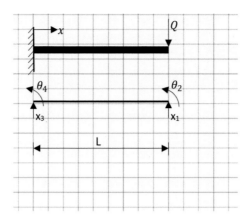

Figure 4.1: Cantilever beam displacement is the optimal solution of minimizing the potential energy function.

the two degrees of freedom at the clamped end vanish; that is: $x_3 = \theta_4 = 0$. Let $\xi = x/L$; the vertical displacement $s(\xi)$ can be computed in terms of the end displacements as follows:

$$s(\xi) = \left[3\xi^2 - 2\xi^3 L\left(\xi^3 - \xi^2\right)\right] \begin{bmatrix} x_1 \\ \theta_2 \end{bmatrix} \tag{4.4}$$

The deflection along the beam can then be found by minimizing its potential energy; the potential energy can be expressed as:

$$
\begin{aligned}
V &= \frac{EI}{2L^3} \int_0^L \left(\frac{d^2 s}{d\xi^2}\right)^2 d\xi + Qx_1 \\
&= \frac{EI}{2L^3}\left(12x_1^2 + 4\theta_2^2 L^2 - 12x_1\theta_2 L\right) + Qx_1 \tag{4.5}
\end{aligned}
$$

where E is the Young's modulus and I is the beam cross section moment of inertia. For a simpler form of Eq. (4.5), let $f = 2VL^3/EI$, $x_1 = x_1$, $x_2 = \theta_2 L$, $QL^3/EI = 1$, and $\vec{x}^T = [x_1 \, x_2]$. Then the objective function that we need to minimize takes the form:

$$
\begin{aligned}
f &= 12x_1^2 + 4x_2^2 - 12x_1x_2 + 2x_1 \tag{4.6} \\
&= \frac{1}{2}\vec{x}^T \begin{bmatrix} 24 & -12 \\ -12 & 8 \end{bmatrix} \vec{x} + \begin{bmatrix} 2 & 0 \end{bmatrix} \vec{x} \tag{4.7}
\end{aligned}
$$

Solve the above optimization problem to find the tip displacement using the Pattern Search Algorithm. Assume the initial guess for the solution be $\vec{x}^0 = [-1 \ -2]^T$.

Solution:

Pattern Search Algorithm: Since we have two variables, then each cycle in the Pattern Search Algorithm was three iterations. At $\vec{x}^0$, the value of the objective function is $f(\vec{x}^0) = 2$. The first search direction is $\vec{s}^0 = [1\ 0]^T$. Hence the next guess can be expressed as function of the search step size α^0 as follows:

$$\vec{x}^1 = \vec{x}^0 + \alpha^0 \vec{s}^0 = \begin{bmatrix} -1 \\ -2 \end{bmatrix} + \alpha^1 \begin{bmatrix} 1 \\ 0 \end{bmatrix} \tag{4.8}$$

The objective function at the new design $\vec{x}^1$ is a function of α^0:

$$\begin{aligned} f(\alpha^0) &= 12\left(\alpha^0 - 1\right)^2 + 4\left(-2\right)^2 + 24\left(\alpha^0 - 1\right) + 2\left(\alpha^0 - 1\right) \\ &= 12(\alpha^0)^2 + 2\alpha^0 + 2 \end{aligned} \tag{4.9}$$

The function $f(\alpha^0)$ is quadratic in α^0 and its minimum point can be obtained directly: $\alpha^0 = -2/24 = -0.0833$. Hence the new design is $\vec{x}^1 = [-13/12\ -2]^T$ and the new objective function value is $f = 1.9167$. This completes the first iteration.

The second iteration has the search direction $\vec{s}^1 = [0\ 1]^T$. Hence the next guess can be expressed as function of the search step size α^1 as follows:

$$\vec{x}^2 = \vec{x}^1 + \alpha^1 \vec{s}^1 = \begin{bmatrix} -13/12 \\ -2 \end{bmatrix} + \alpha^1 \begin{bmatrix} 0 \\ 1 \end{bmatrix} \tag{4.10}$$

The objective function at the new design $\vec{x}^2$ is a function of α^1:

$$\begin{aligned} f(\alpha^1) &= 12\left(-13/12\right)^2 + 4\left(\alpha^1 - 2\right)^2 + 13\left(\alpha^1 - 2\right) - 13/6 \\ &= 4(\alpha^1)^2 - 3\alpha^1 + 1.9166 \end{aligned} \tag{4.11}$$

The function $f(\alpha^1)$ is quadratic in α^1 and its minimum point can be obtained directly: $\alpha^1 = 3/8 = 0.375$. Hence the new design is $\vec{x}^2 = [-13/12\ -13/8]^T$ and the new objective function value is $f = 1.35416$. This completes the second iteration.

The third iteration (pattern search iteration) has a search direction $\vec{s}^p = \alpha^1 \vec{s}^1 + \alpha^0 \vec{x}^0 = [-1/12\ 3/8]^T$. Hence the next guess can be expressed as function of the search step size α^p as follows:

$$\vec{x}^3 = \vec{x}^2 + \alpha^p \vec{s}^p \tag{4.12}$$

The objective function at the new design $\vec{x}^3$ is also a quadratic function of α^p, for which the optimal search step is $\alpha^p = 40/49$. Hence the new design is $\vec{x}^3 = [-157/147\ -83/49]^T$ and the new objective function value is $f = 1.319727$. This completes the third iteration and the first cycle.

The above process repeats until a stopping criterion is satisfied. **Table 4.1** shows the numerical results for all the iterations. The algorithms was stopped after 19 iterations due no change in the design variables x_1 and x_2. Hence the solution is $\vec{x} = [-0.3334\ -0.5001]^T$.

Table 4.1: Cantilever Beam Deflection using Pattern Search Algorithm.

iteration	x_1	x_2	f	$|\alpha|$
1	-1.0000	-2.0000	2.0000	0
2	-1.0833	-2.0000	1.9167	0.0833
3	-1.0833	-1.6254	1.3540	0.3750
4	-1.0681	-1.6938	1.3199	0.1829
5	-0.9307	-1.6938	1.0918	0.1378
6	-0.9307	-1.3965	0.7372	0.2973
7	-0.5541	-0.5818	0.0617	2.7411
8	-0.3744	-0.5818	-0.3266	0.1797
9	-0.3744	-0.5619	-0.3283	0.0199
10	-0.3631	-0.5606	-0.3297	0.0628
11	-0.3635	-0.5606	-0.3297	0.0005
12	-0.3635	-0.5455	-0.3306	0.0152
13	-0.3635	-0.5460	-0.3306	0.0315
14	-0.3564	-0.5460	-0.3312	0.0072
15	-0.3564	-0.5351	-0.3317	0.0113
16	-0.3333	-0.5002	-0.3333	3.2111
17	-0.3334	-0.5002	-0.3333	0.0001
18	-0.3334	-0.5001	-0.3333	0.0001
19	-0.3334	-0.5001	-0.3333	0.1497

4.1.3   Powell's Method

The Powell's Method is similar to the Pattern Search Method, except that the first n search directions in a cycle are copied from the last n iterations in the previous cycle. This method is very popular due to its convergence characteristics. Powell's Method would require only n cycles to converge to the optimal solution when optimizing a quadratic function of n variables.

4.2   Gradient-Based Optimization

No gradient information was needed in any of the methods discussed in **Section** 4.1. In some optimization problems, it is possible to compute the gradient of the objective function, and this information can be used to guide the optimizer for more efficient optimization. Similar to the non-gradient optimization methods, the gradient-based methods in general are iterative methods in which the same update Eq. (4.3) is used. The step size α is computed in a similar way to non-gradient methods. The search direction vector $\vec{s}$, however, is computed based on gradient information when using gradient-based methods. This section presents some of the most popular gradient-based optimization methods.

4.2.1 Steepest Descent Method

The search direction vector at an iterate k is set to be the negative of the objective function gradient at that iterate; that is:

$$\vec{s}^k = -\vec{\nabla} f(\vec{x}^k) \tag{4.13}$$

4.2.2 Conjugate Gradient Method

Consider a quadratic function $f(\vec{x})$ of n variables; f can be expressed in matrix form as:

$$f(\vec{x}) = \frac{1}{2}\vec{x}^T[Q]\vec{x} + \vec{b}^T\vec{x} \tag{4.14}$$

In **Section** 1.1.3.2, it was shown that it is possible to construct n Q-conjugate (Q-orthogonal) directions $\vec{d}^i, i = 1, \cdots n$. Lets express $\vec{x}$ as a linear combination of $\vec{d}^i, \forall i$; that is:

$$\vec{x} = \sum_{i=1}^{n} \alpha^i \vec{d}^i \tag{4.15}$$

Hence we can write the function $f(\vec{x}$ as:

$$f(\vec{x}) = \frac{1}{2}\overbrace{\left(\sum_{i=1}^{n} \alpha^i \vec{d}^i\right)^T [Q] \left(\sum_{i=1}^{n} \alpha^i \vec{d}^i\right)}^{I} + \sum_{i=1}^{n} \alpha^i \vec{b}^T \vec{d}^i \tag{4.16}$$

The term I can be expanded as follow:

$$\begin{aligned}
I &= \frac{1}{2}\left(\alpha^1 \vec{d}^{1,T} + \alpha^2 \vec{d}^{2,T} \cdots \alpha^n \vec{d}^{n,T}\right)[Q]\left(\alpha^1 \vec{d}^1 + \alpha^2 \vec{d}^2 \cdots \alpha^n \vec{d}^n\right) \\
&= \frac{1}{2}\left((\alpha^1)^2 \vec{d}^{1,T}[Q]\vec{d}^1 + (\alpha^2)^2 \vec{d}^{2,T}[Q]\vec{d}^2 \cdots (\alpha^n)^2 \vec{d}^{n,T}[Q]\vec{d}^n\right) \\
&= \frac{1}{2}\sum_{i=1}^{n}(\alpha^i)^2 \vec{d}^{i,T}[Q]\vec{d}^i = \frac{1}{2}\sum_{i=1}^{n}(\alpha^i)^2 \| \vec{d}^i \|_Q^2
\end{aligned} \tag{4.17}$$

Note that it was possible to obtain the final result in Eq. (4.17) because $\vec{d}^{i,T}[Q]\vec{d}^j = 0, \forall i \neq j$ (the vectors $\vec{d}^i$ and $\vec{d}^j$ are Q-orthogonal.) Hence we can express the function $f(\vec{x})$ as:

$$f(\vec{x}) = \sum_{i=1}^{n}\left\{\frac{1}{2}(\alpha^i)^2 \| \vec{d}^i \|_Q^2 + \alpha^i \vec{b}^T \vec{d}^i\right\} = \sum_{i=1}^{n} \kappa_i(\alpha^i) \tag{4.18}$$

The result obtained in Eq. (4.18) is significant; it shows that $f(\vec{x})$ is separable in α; that is $f(\vec{x})$ is a summation of terms each of them is a function of its own variables. Hence when minimizing $f(\vec{x})$ it is possible to minimize each term independently. In other words, we can write:

$$\underset{\vec{\alpha}}{\text{Min}} \, f(\vec{x}) = \sum_{i=1}^{n} \underset{\alpha^i}{\text{Min}} \, \kappa_i(\alpha^i) \tag{4.19}$$

So, we converted the optimization of one function of n variables to the optimization of n functions and each of them is of one variable (a one-dimensional problem).

The Conjugate Directions Method for optimization uses Eq. (4.18), in a recursive manner, starting with optimizing $\kappa_1(\alpha^1)$, then $\kappa_2(\alpha^2)$, and so on. In each iteration, the optimization of the variable α can be carried out numerically using methods such as the Golden Section Method described in **Section** 3.2. The solution in Eq. (4.16) can be constructed recursively and hence the update equation for the solution can be written in the following form:

$$\vec{x}^{k+1} = \vec{x}^k + \alpha^k \vec{d}^k \tag{4.20}$$

It is also possible to assume that $\vec{x} = \vec{x}^0 + \sum_{i=1}^{n} \alpha^i \vec{d}^i$, and we can show that the above conclusion is still valid; that is we can solve the optimization problem by solving n 1-Dimensional optimization problems.

Expanding Manifold Property

Define the subspace $G^k = \left\{ \vec{x} = \vec{x}^0 + \sum_{i=1}^{k} \alpha^i \vec{d}^i \right\}$ after we have performed k steps of Conjugate Directions Method ($k < n$). Then we can show that:

$$\vec{x}^k = \underset{\vec{x} \in G^k}{\text{argmin}} f(\vec{x}) \tag{4.21}$$

At any step (iteration) k, the current iterate $\vec{x}^k$ is the optimal solution for the function $f(\vec{x})$ in the subspace G^k. If we define the vector $\alpha \in \mathbb{R}^k, k < n$, and $[D] = [\vec{d}^1 | \vec{d}^2, \cdots \vec{d}^k]$ then we can write:

$$\vec{x}^{k+1} = \vec{x}^0 + [D]\vec{\alpha} \tag{4.22}$$

Hence,

$$\vec{\alpha}^* = \underset{\vec{\alpha}}{\text{Min}} \, f(\vec{x}) = f(\vec{x}^0 + [D]\vec{\alpha}) \underset{\vec{\alpha}}{\kappa_i(\alpha^i)} \tag{4.23}$$

where $\vec{\alpha}^*$ is the optimal α. Let $\vec{\nabla}_\alpha f$ be the gradient of f over α. Then the necessary condition for optimality is:

$$\begin{aligned} \vec{\nabla}_\alpha f &= \vec{0} \\ \therefore [D]^T \vec{\nabla}_x f(\vec{x}^*) &= \vec{0} \end{aligned} \tag{4.24}$$

Expanding Eq. (4.24), we get:

$$\vec{d}^{1,T}\vec{\nabla}_x f(\vec{x}^*) = 0$$
$$\vec{d}^{2,T}\vec{\nabla}_x f(\vec{x}^*) = 0$$
$$\vdots$$
$$\vec{d}^{k,T}\vec{\nabla}_x f(\vec{x}^*) = 0$$

(4.25)

The result in (4.25) shows that at an iteration k, the gradient of the function f at the current iterate $\vec{x}^*$ is orthogonal to all previous directions $\vec{d}^i, \forall i = 1, \cdots, k$.

Using the Gradient Information

In the above development, we constructed the Q-orthogonal vectors $\vec{d}^i, \forall i = 1, \cdots, n$. As discussed in **Section** 1.1.3.2, it is possible to construct the $\vec{d}$ vectors given a set of linearly independent vectors in $\mathbb{R}^n$. These vectors are here selected to be the negatives of the gradient vectors at each iteration as detailed below. Consider the minimization of the quadratic function $f(\vec{x}) = \frac{1}{2}\vec{x}^T[Q]\vec{x} + \vec{b}^T\vec{x} + c$, then the gradient of f at $\vec{x}^k$ is:

$$\vec{g}^k = \vec{\nabla} f(\vec{x}^k) = [Q]\vec{x}^k + \vec{b}$$

(4.26)

In such case, at each new iterate computed using Eq. (4.20), the gradient vector $\vec{g}^{k+1}$ is computed and used to compute the $\vec{d}^{k+1}$ in a Gram-Schmidt Q-orthogonalization process (see **Section** 1.1.3.2) as follows:

$$\vec{d}^0 = -\vec{g}^0 = -\vec{\nabla} f(\vec{x}^k) = [Q]\vec{x}^k - \vec{b}$$
$$\vdots$$
$$\vec{d}^{k+1} = -\vec{g}^{k+1} + \sum_{i=1}^{k} \frac{\left(\vec{g}^{k+1}, \vec{d}^i\right)_Q}{\|\vec{d}^i\|_Q^2} \vec{d}^i$$

(4.27)

Note that in evaluating the search step α in Eq. (4.20), no numerical approach is needed; an exact line search can be performed to find the optimal α. The above process for optimizing the function f is the Conjugate Gradient Method.

Simplifications

Equation (4.27) can be written as:

$$\vec{d}^{k+1} = -\vec{g}^{k+1} + \sum_{i=1}^{k} \frac{\vec{g}^{k+1,T}[Q]\vec{d}^i}{\vec{d}^{i,T}[Q]\vec{d}^i} \vec{d}^i$$

(4.28)

From the update Eq. (4.29), we can solve for $\vec{d}^k$:

$$\vec{d}^k = \frac{1}{\alpha^k}\left(\vec{x}^{k+1} - \vec{x}^k\right) \tag{4.29}$$

$$\therefore [Q]\vec{d}^k = \frac{1}{\alpha^k}[Q]\left(\vec{x}^{k+1} - \vec{x}^k\right) \tag{4.30}$$

Using Eq. (4.26), then $[Q]\left(\vec{x}^{k+1} - \vec{x}^k\right) = \vec{g}^{k+1} - \vec{g}^k$. Hence,

$$[Q]\vec{d}^k = \frac{1}{\alpha^k}\left(\vec{g}^{k+1} - \vec{g}^k\right) \tag{4.31}$$

Substituting this result into Eq. (4.28), we get:

$$\vec{d}^{k+1} = -\vec{g}^{k+1} + \sum_{i=1}^{k}\frac{\vec{g}^{k+1,T}\left(\vec{g}^{i+1} - \vec{g}^i\right)}{\vec{d}^{i,T}\left(\vec{g}^{i+1} - \vec{g}^i\right)}\vec{d}^i \tag{4.32}$$

Note that the summation term in Eq. (4.32) is a scalar, call it θ^k (summation of scalar terms up to k). The expanding manifold property discussed above showed that $\vec{g}^{k+1}$ is orthogonal to all $\vec{d}^0, \vec{d}^1, \cdots, \vec{d}^k$, and also $\vec{g}^{k+1}$ is orthogonal to all $\vec{g}^0, \vec{g}^1, \cdots, \vec{g}^k$ (to see this, consider a 3-dimensional space $\mathbb{R}^3$, the $\vec{g}^2$) is orthogonal to $\vec{d}^0$ and $\vec{d}^1$. The vector $\vec{g}^0$ is $-\vec{d}^0$, hence $\vec{g}^2$ is also orthogonal to $\vec{g}^0$. From Eq. (4.32), we can see that the vector $\vec{g}^1$ is in the same plane as the vectors $\vec{d}^0$ and $\vec{d}^1$; since $\vec{g}^2$ is orthogonal to both $\vec{d}^0$ and $\vec{d}^1$ then $\vec{g}^2$ is orthogonal $\vec{g}^1$. Hence, $\vec{g}^2$ is orthogonal to both $\vec{g}^1$ and $\vec{g}^0$. As a result of this orthogonality, all terms in the summation in Eq. (4.32) vanished except the last term when $i = k$. Hence we can rewrite Eq. (4.32) as:

$$\vec{d}^{k+1} = -\vec{g}^{k+1} + \frac{\vec{g}^{k+1,T}\left(\vec{g}^{k+1} - \vec{g}^k\right)}{\vec{d}^{k,T}\left(\vec{g}^{k+1} - \vec{g}^k\right)}\vec{d}^k \tag{4.33}$$

Note further that in the denominator of Eq. (4.33), the vector $\vec{g}^{k+1}$ is orthogonal to $\vec{d}^{k,T}$, and hence their scalar product vanished. The remaining term in the denominator is $-\vec{d}^{k,T}\vec{g}^k = -(-\vec{g}^k + \theta^{k-1}\vec{d}^{k-1})^T\vec{g}^k = \vec{g}^{k,T}\vec{g}^k$ ($\vec{g}^k$ *isorthogonaltod*$^{k-1}$).

Hence we can update the search direction using the equation:

$$\vec{d}^{k+1} = -\vec{g}^{k+1} + \frac{\vec{g}^{k+1,T}\left(\vec{g}^{k+1} - \vec{g}^k\right)}{\parallel \vec{g}^k \parallel_2^2}\vec{d}^k \tag{4.34}$$

Using Eq. (4.34) to compute the search directions is known as the Polak-Ribiere Method.

The numerator of Eq. (4.34) can be simplified further by realizing that $\vec{g}^{k+1}$ is orthogonal to $\vec{g}^k$, and hence their scalar product vanishes. Hence we can write:

$$\vec{d}^{k+1} = -\vec{g}^{k+1} + \frac{\| \vec{g}^{k+1} \|_2^2}{\| \vec{g}^k \|_2^2} \vec{d}^k \tag{4.35}$$

Using Eq. (4.35) to compute the search directions is known as the Fletcher-Reeves Method.

Both the Fletcher-Reeves and the Polak-Ribiere methods are Conjugate Gradient methods; both are equivalent when optimizing a quadratic function as can be seen from the above details. In the more general case of non-quadratic objective function, the two methods behave differently. **Algorithm** 4.1 shows the Conjugate Gradient Algorithm.

Algorithm 4.1 Conjugate Gradient Algorithm

Select an initial guess for the solution $\vec{x}^0$, and set k = 0
Compte $\vec{g}^0 = \vec{\nabla} f(\vec{x}^0)$
Compute $\vec{d}^0 = -\vec{g}^0$
while a termination condition is not satisfied **do**
    Compute $\vec{x}^{k+1}(\alpha^k) = \vec{x}^k + \alpha^k \vec{d}^k$
    Find the optimal α^k that optimizes $f(\alpha^k) = f(\vec{x}^{k+1})$
    Compute the new point $\vec{x}^{k+1}$ and the new gradient $\vec{g}^{k+1} = \vec{\nabla} f(\vec{x}^{k+1})$
    Use the Fletcher-Reeves or the Polak-Ribiere methods to compute θ^k
    Compute the new search direction $\vec{d}^{k+1} = -\vec{g}^{k+1} + \theta^k \vec{d}^k$.
    $k = k+1$
end while

4.2.3 Variable Metric Methods

At any iteration k, the Conjugate Gradient methods use gradient information from one previous iteration. It makes sense that algorithms that use all the history of gradients (they are already computed) would be more efficient. The category of methods that do that is called Variable Metric Methods. The *Metric* matrix is the matrix that collects the gradients computed over all the previous iterations. The Variable Metric Methods stand on a solid theoretical foundation and they have practical convergence properties.

Section 4.3 presents second order methods that need the Hessian matrix for optimization; they have quadratic convergence characteristics. The Variable Metric Methods behave like a second order method. Yet the Hessian matrix need not to be computed. In fact, the Hessian matrix can be a by product from a Variable Metric Method as will be seen below. The Variable Metric Methods are also called quasi-Newton Methods. As before, they update the solution at each iteration using the following equation:

$$\vec{x}^{k+1} = \vec{x}^k + \alpha^k \vec{s}^k \tag{4.36}$$

where the search direction $\vec{s}$ is computed as a function of the gradients.

Davidon-Fletcher-Powell (DFP) Method

The search direction is updated at an iteration k using the equation:

$$\vec{s}^k = -[A^k]\vec{\nabla} f(\vec{x}^k) \tag{4.37}$$

where the matrix $[A]$ is initialized by the user and is updated each iteration using the process outlined in **Algorithm** 4.2.

Algorithm 4.2 Davidon-Fletcher-Powell (DFP) Algorithm

Select an initial guess for the solution $\vec{x}^0$, and the matrix $[A^0]$. Set $k = 0$
Compte $\vec{g}^0 = \vec{\nabla} f(\vec{x}^0)$
Compute $\vec{s}^0 = -[A^0]\vec{g}^0$
while a termination condition is not satisfied **do**
    Compute $\vec{x}^{k+1}(\alpha^k) = \vec{x}^k + \alpha^k \vec{s}^k$
    Find the optimal α^k that optimizes $f(\alpha^k) = f(\vec{x}^{k+1})$
    Compute the new solution $\vec{x}^{k+1}$, and the new gradient $\vec{g}^{k+1} = \vec{\nabla} f(\vec{x}^{k+1})$
    Compute the $\vec{\Delta x} = \vec{x}^{k+1} - \vec{x}^k$
    Compute the vector $\vec{\Delta g} = \vec{g}^{k+1} - \vec{g}^k$
    Compute the matrix $[D] = \dfrac{\vec{\Delta x}\vec{\Delta x}^T}{\vec{\Delta x}^T \vec{\Delta g}}$
    Compute the matrix $[C] = -\dfrac{[A^k]\vec{\Delta g}\vec{\Delta g}^T[A^k]^T}{\vec{\Delta g}^T[A^k]\vec{\Delta g}}$
    Update the current $[A]$ matrix: $[A^{k+1}] = [A^k] + [C] + [D]$
    $k = k + 1$
end while

The $[A]$ matrix converges to the inverse of the Hessian matrix in the DFP method.

Broydon-Fletcher-Goldfarb-Shanno (BFGS) Method

The search direction is updated at an iteration k using the equation:

$$[A^k]\vec{s}^k = -\vec{\nabla} f(\vec{x}^k) \tag{4.38}$$

where the matrix $[A]$ is initialized by the user and is updated at each iteration using the process outlined in **Algorithm** 4.3. The $[A]$ matrix converges to the Hessian matrix in the BFGS method.

Algorithm 4.3 Broydon-Fletcher-goldfarb-Shanno (BFGS) Algorithm

---

Select an initial guess for the solution $\vec{x}^0$, and the matrix $[A^0]$. Set $k = 0$
Compte $\vec{g}^0 = \vec{\nabla} f(\vec{x}^0)$
Compute $\vec{s}^0 = -[A^0]^{-1}\vec{g}^0$
while a termination condition is not satisfied **do**
    Compute $\vec{x}^{k+1}(\alpha^k) = \vec{x}^k + \alpha^k \vec{s}^k$
    Find the optimal α^k that optimizes $f(\alpha^k) = f(\vec{x}^{k+1})$
    Compute the new solution $\vec{x}^{k+1}$, and the new gradient $\vec{g}^{k+1} = \vec{\nabla} f(\vec{x}^{k+1})$
    Compute the $\vec{\Delta x} = \vec{x}^{k+1} - \vec{x}^k$
    Compute the vector $\vec{\Delta g} = \vec{g}^{k+1} - \vec{g}^k$
    Compute the matrix $[D] = \dfrac{\vec{\Delta g}\vec{\Delta g}^T}{\vec{\Delta g}^T \vec{\Delta x}}$
    Compute the matrix $[C] = -\dfrac{\vec{g}^k \vec{g}^{k,T}}{\vec{g}^{k,T}\vec{s}^k}$
    Update the current $[A]$ matrix: $[A^{k+1}] = [A^k] + [C] + [D]$
    $k = k + 1$
end while

4.3 Second Order Methods

The update equation in a second order method is the same as the update equation in the Variable Metric Methods, Eq. (4.36). The update matrix $[A]$ in the DFP and BFGS algorithms (both are first order methods) is computed as function of the gradient information. In the second order method, however, the update matrix is the Hessian matrix. That is the search direction is updated as follows:

$$\vec{s}^k = -[H(\vec{x}^k)]\vec{\nabla} f(\vec{x}^k) \tag{4.39}$$

Clearly, this method requires computing the Hessian matrix; hence it may not be practical in many applications due to the computational cost associated with computing the Hessian matrix. The second order method is outlined in **Algorithm** 4.4.

Example 4.2. Find the minimum of the function $f(x_1, x_2)$, where

$$f(x_1, x_2) = 0.5x_1^2 + x_2^2 \tag{4.40}$$

Algorithm 4.4 Algorithm for The Second Order Method

Select an initial guess for the solution $\vec{x}^0$, Set $k = 0$
while a termination condition is not satisfied **do**
    Compte $\vec{g}^k = \vec{\nabla} f(\vec{x}^k)$
    Compte the Hessian matrix $[H(\vec{x}^k)]$
    Compute $\vec{s}^k = -[H]^{-1} \vec{g}^0$
    Compute $\vec{x}^{k+1}(\alpha^k) = \vec{x}^k + \alpha^k \vec{s}^k$
    Find the optimal α^k that optimizes $f(\alpha^k) = f(\vec{x}^{k+1})$
    Compute the new solution $\vec{x}^{k+1}$, and the new gradient $\vec{g}^{k+1} = \vec{\nabla} f(\vec{x}^{k+1})$
    $k = k + 1$
end while

Solution: The gradient of the function $f(x_1, x_2)$ is:

$$\vec{\nabla} f(x_1, x_2) = \begin{bmatrix} x_1 \\ 2x_2 \end{bmatrix} \tag{4.41}$$

The Hessian of the function $f(x_1, x_2)$ is:

$$[H(x_1, x_2)] = \begin{bmatrix} 1 & 0 \\ 0 & 2 \end{bmatrix} \tag{4.42}$$

Assume an initial guess of $x_1 = -2$, and $x_2 = -3$. The fist update is:

$$\vec{x}^1 = \begin{bmatrix} -2 \\ -3 \end{bmatrix} - \begin{bmatrix} 1 & 0 \\ 0 & 1/2 \end{bmatrix} \begin{bmatrix} -2 \\ -6 \end{bmatrix} = \begin{bmatrix} 0 \\ 0 \end{bmatrix} \tag{4.43}$$

The solution is obtained after one iteration for this second order function. Try using any other initial guess.

Chapter 5

Numerical Algorithms for Constrained Optimization

In general, constrained optimization problems are more challenging to solve, compared to the unconstrained optimization problems. The general form of the optimization problem is:

$$\text{minimize: } f(\vec{x})$$
$$\text{subject to: } h_i(\vec{x}) = 0, i = 1 \cdots l. \tag{5.1}$$
$$g_j(\vec{x}) \leq 0, j = 1 \cdots m.$$

where $\vec{x} \in \mathbb{R}^n$, $\mathbb{R}^n$ is the n-dimensional Euclidian space, $h_i : \mathbb{R}^n \longrightarrow \mathbb{R}$, $g_j : \mathbb{R}^n \longrightarrow \mathbb{R}$, and $f : \mathbb{R}^n \longrightarrow \mathbb{R}$, $\forall i, j$.

There are several numerical algorithms that can be used to search for the optimal solution of a function in the presence of constraints. These methods can be categorized, in general, into two categories. The first category is called Direct Methods, and it includes the methods that approximate the objective and constraint functions, to simplify the problem. The second category of methods is called Indirect Methods, and it includes methods that convert the constrained optimization problem to a sequence of unconstrained optimization problems, to leverage the unconstrained optimization algorithms for solving the more complex constrained problems. In both types of methods, the numerical algorithm usually starts with an initial guess $\vec{x}^0$, and then updates this initial guess over subsequent iterations using the update equation:

$$\vec{x}^{k+1} = \vec{x}^k + \alpha^k \vec{s}^k \tag{5.2}$$

The search direction $\vec{s}^k$ and step α^k will vary from one method to another. This chapter summarizes some of the methods in each category. We start with the Indirect Methods.

5.1 Indirect Methods

As mentioned above, Indirect Methods convert the constrained optimization problem to a sequence of unconstrained optimization problems. This is achieved by appending the objective function with "penalty functions" that penalizes the violation of the constraints. In other words, we search for the optimal solution for the optimization problem (5.1) by solving another unconstrained optimization problem, that is of the form:

$$\text{minimize: } F(\vec{x}, w_h, w_g) = f(\vec{x}) + P(\vec{x}, w_h, w_g)$$
$$\text{Subject to the boundaries: } \vec{x}^l \leq \vec{x} \leq \vec{x}^u \tag{5.3}$$

where $x \in \mathbb{R}^n$, w_h and w_g are penalty multipliers associated with the equality and inequality constraints, respectively. The equality and inequality constraint functions are all included in the penalty functions, as detailed in the sections below. The form of the penalty function $P(\vec{x}, w_h, w_g)$ and how it includes the constraints functions depends on the specific method.

In general the selection of the penalty parameters values determine the relative weight given to optimality (minimizing $f(\vec{x})$) versus feasibility of the obtained solution (satisfying the constraints.) The unconstrained problem (5.3) is solved several times, sequentially, using different selections of the penalty parameters in each iteration. At each iteration in this sequence, an unconstrained optimization algorithm is used to search for the optimal solution of (5.3), where the initial guess is the solution obtained from the previous iteration. An algorithm for indirect methods is given later in this chapter in **Algorithm** 5.1. Some of the main Indirect Methods are summarized below.

5.1.1 Barrier Methods

The Barrier methods construct the augmented function of the associated unconstrained optimization problem in such a way that the solution at any iteration does not leave the feasible domain. These methods are also called The Interior Penalty Function Methods for that same reason - the solutions obtained in subsequent iterations are inside the feasible domain. The fact that the solutions are always in the feasible "domain" implies that this type of problem would not work for problems that have equality constraints, since in the presence of equality constraints there might be no domain to iterate on. The Barrier methods works for problems that have only the inequality constraints. In Barrier methods, the penalty function $P(\vec{x}, w_g)$ is continuous and nonnegative over the feasible domain, and it approaches infinity as the solution approaches the boundary of the feasible domain. There are several candidates for the penalty function, two of them are:

$$P(\vec{x}, w_g) = -w_g \sum_{i=1}^{m} log\left(-g_i(\vec{x})\right) \tag{5.4}$$

$$P(\vec{x}, w_g) = -w_g \sum_{i=1}^{m} \frac{1}{g_i(\vec{x})} \tag{5.5}$$

When using the Barrier methods, we start the iterations with high values for the penalty parameter w_g, and then we decrease it over subsequent iterations. This basically means that in the initial iterations we give high weight for moving the solutions away from the feasible domain boundaries toward the interior of it. In subsequent iterations, we give increasing weight to the optimality of the solution.

Example 5.1. Find the minimum of $f(x)$ using the Barrier method.

$$\begin{aligned} \text{minimize: } f(x) &= 5x^2 \\ \text{subject to: } 3 - x &\leq 0 \end{aligned} \tag{5.6}$$

Solution: The associated unconstrained optimization problem can be written as:

$$\text{minimize: } F(x, w_g) = 5x^2 - w_g log(x - 3) \tag{5.7}$$

The feasible domain is $x > 3$ in order for the *log* function to be valid. The derivative of F is:

$$\frac{dF}{dx} = 10x - \frac{w_g}{x - 3} \tag{5.8}$$

To find the minimum of F, we set the derivative $\dfrac{dF}{dx} = 0$. Hence,

$$10x(x - 3) - w_g = 0$$

$$\therefore x = 1.5 \pm \frac{\sqrt{900 + 40w_g}}{20} \tag{5.9}$$

We can find the optimal solution by taking the limit as $w_g \to 0$:

$$x = \lim_{w_g \to 0} 1.5 \pm \frac{\sqrt{900 + 40w_g}}{20} = 1.5 \pm 1.5 \tag{5.10}$$

The solution $x = 0$ is excluded because it is outside the feasible domain ($x \geqslant 3$). Hence the optimal solution is $x = 3$.

Equation (5.9) can be used to plot the obtained solution for different values of the penalty parameter. **Figure** 5.1 shows the values of x for a range of $w_g = 10^3 \cdots 10^{-8}$, where it is clear that x converges to the optimal solution as w_g decreases. It is observed from **Fig.** 5.1 that all values of x for different values of w_g are in the feasible domain $x > 3$.

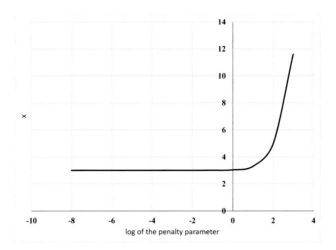

x

log of the penalty parameter

Figure 5.1: Using Barrier methods, as the penalty paramter decreases, the solution converges to the optimal value.

In the above simple example, it was possible to express the optimal solution in terms of the penalty parameter w_g, and hence it was possible to take the limit and get the optimal solution. In more complex problems, this is not possible. Rather, we select a value for penalty parameter w_g; then we numerically solve the unconstrained optimization problem (5.3). The obtained solution for $\vec{x}$ is then used as initial guess, and we solve the unconstrained optimization problem (5.3) using a new value for w_g. This iterative process repeats until a convergence is achieved. The penalty parameter w_g is updated in every iteration using a simple update equation:

$$w_g^{k+1} = w_g^k C_g \tag{5.11}$$

where k is the iteration index, and C_g is a constant multiplier whose value depends on the specific problem. An algorithm for indirect methods is given in **Algorithm** 5.1 which applies to the Barrier methods.

5.1.2   *Exterior Penalty Function Methods*

As can be inferred from the name, the Exterior Penalty Function Methods start with an initial guess outside of the feasible domain, and approaches the optimal solution from outside. Because the iterations approach from outside the feasible, this methods finds the extremals near the boundaries of the feasible domain. The Exterior Penalty Function methods can handle both equality and inequality constraints.

In exterior penalty function methods, the penalty function may take the general form:

$$P(\vec{x}, w_h, w_g) = w_h \sum_{j=1}^{l} (h_j(\vec{x}))^2 + w_g \sum_{i=1}^{m} (max\{0, g_i(\vec{x})\})^2 \tag{5.12}$$

As can be inferred from Eq. (5.12), as $w_h \to \infty$ and $w_g \to \infty$, the unconstrained optimization problem in (5.3) becomes equivalent to the original constrained optimization problem in (5.1).

The following example illustrates this concept of exterior penalty function methods.

Example 5.2.

$$\text{minimize: } f(x) = 5x^2$$
$$\text{subject to: } 3 - x \leq 0 \tag{5.13}$$

Solution: The associated unconstrained optimization problem can be written as:

$$\text{minimize: } F(x, w_g) = 5x^2 + w_g (max\{0, 3 - x\})^2 \tag{5.14}$$

The augmented function $F(x, w_g)$ can be written as:

$$F(x, w_g) = \begin{cases} 5x^2 + w_g (3 - x)^2, & x < 3 \\ 5x^2, & x \geq 3 \end{cases} \tag{5.15}$$

Hence, we can write:

$$\frac{dF}{dx} = \begin{cases} 10x + 2w_g (3 - x), & x < 3 \\ 10x, & x \geq 3 \end{cases} \tag{5.16}$$

To find the minimum of F, we set the derivative $\frac{dF}{dx} = 0$. Hence,

$$x = \begin{cases} \dfrac{6w_g}{2w_g - 10}, & x < 3 \\ 0, \ (Not feasible, excluded) & x \geq 3 \end{cases} \tag{5.17}$$

Clearly, the solution $x = 0$ is excluded because it is outside the feasible domain $(x \geq 3)$. For $x < 3$, we can find the optimal solution by taking the limit as $w_g \to \infty$:

$$x = \lim_{w_g \to \infty} \frac{6w_g}{2w_g - 10} = 3 \tag{5.18}$$

Hence the optimal solution is $x = 3$. Eq. (5.18) can be used to plot the obtained solution for different values of the penalty parameter. **Figure** 5.2 shows the values of x for a range of $w_g = 1 \cdots 10^{12}$, where it is clear that x converges to the

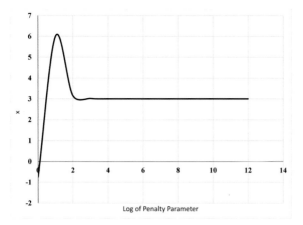

Figure 5.2: As the penalty paramter increases, the solution converges to the optimal value.

optimal solution as w_g increases. It is observed from **Fig.** 5.2 that there is over-shoot that is actually nonfeasible, since $x < 3$.

In general, we cannot just use high values for the penalty parameters because it will result in numerical difficulty in searching for the optimal solution. In other words, giving high values for the penalty parameters, at initial iterations, puts high weight on the feasibility of the solution compared to the optimality of the objective function. In practice, one would start with lower values for the penalty parameters, at initial iterations, and increase them in subsequent iterations. The penalty parameters are updated according the equation:

$$
\begin{aligned}
w_g^{k+1} &= w_g^k C_g \\
w_h^{k+1} &= w_h^k C_h
\end{aligned}
\tag{5.19}
$$

where k is the iteration index, C_g and C_h are constant multipliers whose values depend on the specific problem. The algorithm for indirect methods given in **Algorithm** 5.1 applies also to the Exterior Penalty Function methods.

5.1.3 *Augmented Lagrange Multiplier Method*

The Augmented Lagrange Multiplier method is considered one of the most robust indirect methods. Here, the penalty function $P(\vec{x}, w_g, w_h)$ is defined as:

$$
\begin{aligned}
P(\vec{x}, w_h, w_g, \vec{\lambda}, \vec{\beta}) = & w_h \sum_{j=1}^{l} h_j(\vec{x})^2 + w_g \sum_{i=1}^{m} max\left(-\frac{\beta_i}{2w_g}, g_i(\vec{x})\right)^2 \\
& + \sum_{j=1}^{l} \lambda_j h_j(\vec{x}) + \sum_{i=1}^{m} \beta_i max\left(-\frac{\beta_i}{2w_g}, g_i(\vec{x})\right)
\end{aligned}
\tag{5.20}
$$

where $\vec{\lambda} = [\lambda_1 \cdots \lambda_l]^T$ is a Lagrange multipliers vector associated with the equality constraints, and $\vec{\beta} = [\beta_1 \cdots \beta_m]^T$ is a Lagrange multipliers vector associated with the inequality constraints. The augmented function in this case is:

$$F(\vec{x}, w_h, w_g, \vec{\lambda}, \vec{\beta}) = f(\vec{x}) + P(\vec{x}, w_h, w_g, \vec{\lambda}, \vec{\beta}) \tag{5.21}$$

The Augmented Lagrange Multiplier method works in a similar way to that of the Barrier and Exterior Penalty Function methods. The algorithm starts with initial guess for the solution $\vec{x}^0$, and solves the unconstrained optimization problem to minimize the augmented function F. Any algorithm for unconstrained optimization (such as the BFGS or the DFP algorithms) can be used to solve the unconstrained optimization problem. The obtained solution is used as a new guess $\vec{x}^1$ to replace $\vec{x}^0$, and the unconstrained optimization problem is solved again in this new iteration. The iterations continue until a stopping criterion is satisfied. The penalty parameters w_g and w_h along with the Lagrange multipliers $\vec{\lambda}$ and $\vec{\beta}$ are updated in each iteration. The w_g and w_h are updated in a similar way as that of the Exterior Penalty Function method, given in Eq. (5.19). The Lagrange multipliers are updated according to the equation:

$$\begin{aligned}
\lambda_j^{k+1} &= \lambda_j^k + 2w_h^k h_j(\vec{x}^k), &\quad \forall j = 1 \cdots l \\
\beta_i^{k+1} &= \beta_i^k + 2w_g^k max\left(-\frac{\beta_i^k}{2w_g^k}, g_i(\vec{x}^k)\right), &\quad \forall i = 1 \cdots m
\end{aligned} \tag{5.22}$$

where k is the iteration index.

To get more insight into the method, consider the necessary conditions of optimality discussed in **Chapter** 3. The set of the necessary conditions associated with the inequality constraints are given in Eq. (3.25). The condition in Eq. (3.25) implies that either $\beta_i = 0$ or $g_i(\vec{x}) = 0$, $\forall i = 1 \cdots m$. It can be shown that in the case of $g_i(\vec{x}) = 0$, $\beta_i > 0$. Hence, it is possible to state that a necessary condition for optimality is that $\beta_i \geq 0$. Hence, the optimal solution is expected to be in the regions where β_i is non-negative. The last term in the penalty function in (5.20) penalizes negative values of β_i. In the case when $\beta_i = 0$, the feasibility of the solution necessitates that $g_i(\vec{x}) < 0$. Hence, we can summarize this necessary condition as follows:

$$\begin{aligned}
\beta_i &= 0, g_i(\vec{x}) < 0 \\
&Or \\
\beta_i &> 0, g_i(\vec{x}) = 0
\end{aligned} \tag{5.23}$$

By considering all the possible sign combinations of β_i and $g_i(\vec{x})$, it is easy to show that the second term in the penalty function in (5.20) will result in a value of zero only when the condition (5.23) is satisfied. If the condition (5.23) is not satisfied, the second term in the penalty function in (5.20) will result in a positive value. The Augmented Lagrange Multiplier method uses the necessary conditions of optimality in the construction of the penalty term of the augmented function, and hence it guides the numerical search for the optimal solution based on information learned from the necessary conditions of optimality.

5.1.4 Algorithm for Indirect Methods

Algorithm 5.1 below is an algorithm for general indirect methods such as the three methods described in the above three sections. Overall, the algorithm for any Indirect Method (IM) has two loops. The outer loop is the IM loop, which updates the IM parameters. The IM parameters are penalty parameters and the Lagrange multipliers if applicable. The inner loop has the unconstrained optimization algorithm, which itself is an iterative algorithm, e.g., the DFP and the BFGS algorithms. After each iteration of the IM loop, the convergence is checked. The method is converged if all the constraints $h_j(\vec{x}) = 0$ and $g_i(\vec{x}) \leq 0$ are satisfied, $\forall i, j$, side constraints are satisfied. In the case of Augmented Lagrange Multiplier method, an additional condition needs to be satisfied which is (5.23).

Algorithm 5.1 Algorithm for Indirect Methods

Select an initial guess for the solution $\vec{x}^0$

Select the maximum number of IM loop iterations N_{itr}

Select an initial value for each of the penalty parameters and Lagrange multipliers, as applicable

Select the constant multipliers C_h and C_g

k = 0

while k $\leq N_{itr}$ **do**

    Call the BFGS (or DFP) solver to optimize the augmented function F, and output $\vec{x}^{k+1}$

    if Convergence conditions are satisfied **then**

        STOP

    end if

    if Change in $\vec{\Delta x} = \vec{x}^{k=1} - \vec{x}^k$ is small **then**

        STOP

    end if

    if Change $\Delta F = F(\vec{x}^{k+1}, w_g^{k+1}, w_h^{k+1}, \vec{\lambda}^{k+1}, \vec{\beta}^{k+1}) - F(\vec{x}^k, w_g^k, w_h^k, \vec{\lambda}^k, \vec{\beta}^k)$ is small **then**

        STOP

    end if

    k = k + 1

    Update $w_g^{k+1}, w_h^{k+1}, \vec{\lambda}^{k+1}, \vec{\beta}^{k+1}$ using Eqns. (5.19) and (5.22)

end while

It is noted that the penalty parameters change over subsequent iterations, and hence the augmented function F also changes over subsequent iterations. Hence, in each iteration the unconstrained optimization algorithm is used to solve a different problem of different objective F.

5.2   Direct Methods

In Direct methods, the search for the optimal solution is carried out by sequential approximation of the objective and constraint functions. This approximation enables solving the constraint optimization problem at each iteration. The sequential quadratic programming is one of the widely used Direct methods. It is presented below in **Section** 5.2.3. We first present the sequential linear programming algorithm.

5.2.1   *Sequential Linear Programming*

In Sequential Learning Programming (SLP), each of the functions $f(\vec{x})$, $h_i(\vec{x})$, and $g_j(\vec{x})$ in (5.1) is linearly approximated using Taylor expansion, $\forall i, j$. As a result, we get a linear programming problem, which can be solved using the Simplex method, as discussed in **Chapter** 3. This linear approximation, however, is valid only in a small range around the linearization point. The optimal solution to the original problem (5.1) is unknown, and hence we do not know where to linearize the functions. To search for the optimal solution using the linearized model, we follow an iterative process. First, we start with an initial guess for the solution, and we linearize the problem functions around that guess. The linearized optimization problem around a current guess $\vec{x}^k$ is:

$$\begin{aligned}
\text{minimize: } \overline{f}(\vec{x}) \quad &= f(\vec{x}^k) + \vec{\nabla} f^T(\vec{x}^k)\left(\vec{x} - \vec{x}^k\right) \\
\text{subject to: } \overline{h}_i(\vec{x}^k) \quad &= h_i(\vec{x}^k) + \vec{\nabla} h_i^T(\vec{x}^k)\left(\vec{x} - \vec{x}^k\right) = 0, i = 1 \cdots l. \qquad (5.24) \\
\overline{g}_j(\vec{x}^k) \quad &= g_j(\vec{x}^k) + \vec{\nabla} g_j^T(\vec{x}^k)\left(\vec{x} - \vec{x}^k\right) \leq 0, j = 1 \cdots m.
\end{aligned}$$

We then solve the above linear programming problem for $\vec{x}$. The obtained solution $\vec{x}$ is then used as a new guess $\vec{x}^k$ for the next iteration. The process repeats by linearizing all the functions around the current guess, and solving for a new guess. This process repeats until one of the stopping criteria is met. In addition to satisfying the constraints and side constraints, the stopping criteria are usually:

- ■ if $f(\vec{x}^{k+1}) > f(\vec{x}^k)$, then stop. Or,

- ■ if $\left(\vec{x} - \vec{x}^k\right)^T\left(\vec{x} - \vec{x}^k\right)$ is less than a specified tolerance, then stop.

The sequential linear programming method is not widely used. The linear approximation is valid only in a very small range around the linearization point. This method is not robust.

5.2.2   *Quadratic Programming*

The quadratic programming algorithm searches for the optimal solutions of optimization problems in which the objective function is a quadratic function and all constraint functions are linear. There are indeed several approaches that can be used to solve this type of problem. We start with reviewing some of the characteristics of quadratic functions.

For a quadratic function of one variable $f(y) = ay^2 + by + c$, the gradient is:

$$\nabla f(y) = \frac{df}{dy} = 2ay + b \tag{5.25}$$

The Hessian is a scalar in this case and is given as:

$$[H(f)] = \frac{d^2 f}{dy^2} = 2a \tag{5.26}$$

If we want to:

$$\text{Minimize: } f(y) = ay^2 + by + c \tag{5.27}$$

then the necessary condition for optimality yields:

$$\frac{df}{dy} = 2ay + b = 0 \Longrightarrow y^* = \frac{-b}{2a} \tag{5.28}$$

A quadratic function of n variables can be written in a matrix form as:

$$f(\vec{x}) = \vec{x}^T [A] \vec{x} + \vec{b}^T \vec{x} + c \tag{5.29}$$

For the above quadratic function we can write the gradient vector as:

$$\vec{\nabla} f(\vec{x}) = ([A] + [A]^T) \vec{x} + \vec{b} \tag{5.30}$$

The Hessian matrix is:

$$[H(f)] = [A] + [A]^T \tag{5.31}$$

If we want to:

$$\text{Minimize: } f(\vec{x}) = \vec{x}^T [A] \vec{x} + \vec{b}^T \vec{x} + c \tag{5.32}$$

then the necessary condition for optimality yields:

$$([A] + [A]^T) \vec{x} + \vec{b} = \vec{0} \tag{5.33}$$

If the matrix $([A] + [A]^T)$ is invertible, the extremum point is then:

$$\vec{x}^* = -([A] + [A]^T)^{-1} \vec{b} \tag{5.34}$$

In the remaining of this section we will look at optimization problems in which the objective function $f(\vec{x})$ is quadratic as defined in Eq. (5.29), and all the equality and inequality constrains are linear. This will be presented through examples.

Example 5.3. Use the variable elimination method to find the optimal solution for the following problem:

$$\text{Minimize: } f(x,y) = x^2 + y^2 + xy + y$$
$$\text{Subject to: } h(x,y) \equiv x - y - 1 = 0$$

Solution:
We have only one equality constraint and it is linear. Hence we can use this constraint to reduce the number of variables and eliminate the constraint function as follows. From the constraint in (5.35), we can write:

$$x = 1 + y \tag{5.35}$$

Substituting from Eq. (5.35) in $f(x,y)$ we get:

$$f(y) = 3y^2 + 4y + 1 \tag{5.36}$$

Hence the optimization problem given in (5.35) can be rewritten as an unconstrained optimization problem as follows:

$$\text{Minimize: } f(y) = 3y^2 + 4y + 1 \tag{5.37}$$

Moreover the objective function is quadratic which has a closed form analytic solution to the extreme point given in Eq. (5.28). So the optimal solution is at $y^* = \dfrac{-2}{3}$. Substituting y^* into Eq. (5.35) we get $x^* = \dfrac{1}{3}$.

Example 5.4. Solve the optimization problem given in **Example** 5.3 using the Exterior Penalty Function method.
Solution:
In this problem there is only one equality constraint; hence we introduce a penalty function of only one parameter w_h. The optimization of the augmented function $F(x,y,w_h)$ in this case is then:

$$\text{Minimize: } F(\vec{x}, w_h) = x^2 + y^2 + xy + y + w_h (x - y - 1)^2 \tag{5.38}$$

The function $F(\vec{x}, w_h)$ can be expressed in a matrix form as:

$$F(\vec{x}, w_h) = (1 + w_h)x^2 + (1 + w_h)y^2 + (1 - 2w_h)xy - 2w_h x + (2w_h + 1)y + w_h$$

$$= \vec{x}^T \begin{bmatrix} 1 + w_h & \dfrac{1 - 2w_h}{2} \\ \dfrac{1 - 2w_h}{2} & 1 + w_h \end{bmatrix} \vec{x} + [-2w_h, \; 2w_h + 1]\vec{x}$$

Minimizing the function $F(\vec{x}, w_h)$ is unconstrained optimization, and the objective function is quadratic. The necessary conditions of optimality as given by Eq. (5.34) yields the solution as:

$$\vec{x} = -[H(F)]^{-1}\vec{b} \tag{5.39}$$

where,

$$[H(F)] = \begin{bmatrix} 2(1+w_h) & 1-2w_h \\ 1-2w_h & 2(1+w_h) \end{bmatrix} \tag{5.40}$$

$$\therefore \vec{x} = -\frac{1}{12w_h}\begin{bmatrix} 2(1+w_h) & 2w_h-1 \\ 2w_h-1 & 2(1+w_h) \end{bmatrix}\begin{bmatrix} -2w_h \\ 2w_h+1 \end{bmatrix}$$

$$= -\frac{1}{12w_h}\begin{bmatrix} -4w_h-1 \\ 8w_h+2 \end{bmatrix} \tag{5.41}$$

The Exterior Penalty Function method approaches the optimal solution as w_h gets larger, hence we can write:

$$\vec{x}^* = \lim_{w_h \to \infty} -\frac{1}{12w_h}\begin{bmatrix} -4w_h-1 \\ 8w_h+2 \end{bmatrix} = \lim_{w_h \to \infty} \frac{1}{12}\begin{bmatrix} 4+1/w_h \\ -8-2/w_h \end{bmatrix} = \begin{bmatrix} 1/3 \\ -2/3 \end{bmatrix} \tag{5.42}$$

The above two examples showed how to solve quadratic optimization problems when there is only linear equality constraints. In the presence of linear inequality constraints, the inequality constraint can be handled in a similar way as the equality constrains if methods such as the Exterior Penalty Function or the Augmented Lagrange Multiplier are used. The inequality constraints can be handled more efficiently, however, if we notice that an inequality constraint will only affect the optimal solution if it is active at the optimal solution (i.e., $g_i(\vec{x}^*) = 0$, where $\vec{x}^*$ is the optimal solution;) in such case the active inequality constraints can be handled as equality constraints. To better see that, consider the very simple case of minimizing the quadratic function $f(x) = x^2 + 5$ subject to the constraint $x \leq -2$. In this case, the optimal solution to the unconstrained problem is $x = 0$, but obviously $x = 0$ is outside the feasible domain. One can conclude that the inequality constraint has to be active in this case and the optimal solution can be found if we minimize $f(x)$ subject to the constraint $x = -2$. In a more general case, given the m inequality constraints, at each iterate $\vec{x}^k$, one needs to determine the set of active constraints $J_{ac}(\vec{x}^k) := \{j \in \{1, \cdots, m\} | g_j(\vec{x}^k) = 0\}$, and its compliment (the set of inactive constraints) $J_{in}(\vec{x}^k) := \{1, \cdots, m\} \setminus J_{ac}(\vec{x}^k)$. In this approach for solving quadratic programming problems, the set $J_{ac}(\vec{x}^k)$ are those inequality constraints that are not satisfied at the current iterate $\vec{x}^k$, and they are handled as equality constraints, while the set $J_{in}(\vec{x}^k)$ includes the rest of the inequality constraints, and they can be ignored at the current iterate $\vec{x}^k$. This is further illustrated in the following example.

Example 5.5.

$$\text{Minimize: } f(x,y) = (x-1)^2 + (y-2.5)^2$$
$$\text{Subject to: } h(x,y) \equiv 2y - x - 2 = 0 \qquad (5.43)$$
$$g(x,y) \equiv -x - 2y + 6 \geq 0$$

Solution: Variable elimination can be used to eliminate the equality constraint $h(x,y)$ and the variable x from the objective function f as discussed in **Example** 5.3. Ignoring the inequality constraint for now, the problem reduces to:

$$\text{Minimize: } f(y) = 5y^2 - 17y + 15.25 \qquad (5.44)$$

For the optimization problem (5.44), the optimal solution is $\vec{x} = [1.4, 1.7]^T$. Now we check if the obtained solution satisfies the inequality constraints by direct substitution in $g(\vec{x})$. In this example, the inequality constraint is satisfied and hence the problem is complete. That is $\vec{x}^* = [1.4, 1.7]^T$.

Now consider solving the same problem, but with a different inequality constraint function. The new $g(\vec{x})$ is defined as:

$$g(x,y) \equiv -x - 2y + 1 \geq 0 \qquad (5.45)$$

In this case, the solution $\vec{x} = [1.4, 1.7]^T$ does not satisfy the inequality constraint. Hence $g(\vec{x})$ is active. That is the optimization problem can now be written as:

$$\text{Minimize: } f(x,y) = (x-1)^2 + (y-2.5)^2$$
$$\text{Subject to: } h_1(x,y) \equiv 2y - x - 2 = 0 \qquad (5.46)$$
$$h_2(x,y) \equiv -x - 2y + 1 = 0$$

It is straight forward to show that the optimal solution for this problem is $\vec{x}^* = [-1/2, 3/4]^T$ The quadratic programming method applies only to problems of quadratic objective functions and linear constraint functions. Yet, it is a building block in the more widely used method called Sequential Quadratic Programming. The function quadprog.m in Matlab solves quadratic programming problems.

5.2.3   *Sequential Quadratic Programming*

The Sequential Quadratic Programming (SQP) is an iterative method that solves the general optimization problem in a direct way by approximating the objective and constraint functions. A quadratic function is used to approximate the objective function at the current guess (iterate) for the solution, while a linear function is used to approximate each of the constraint functions at each iterate. Hence the SQP approximates the general nonlinear programming problem to quadratic programming problem at each iteration. The SQP is one of the most effective methods. SQP was first proposed in 1960s by R. Wilson in his PhD dissertation at Harvard University. Since then, the SQP has evolved significantly, based on a rigorous theory, to a class of methods suitable for a wide range of problems.

The standard nonlinear programming problem is given in (5.1). Using the SQP, in each iteration k, the guess $\vec{x}^k$ is used to model the given nonlinear programming problem as a quadratic programming subproblem. This subproblem is solved and the solution is used to compute the new iterate $\vec{x}^{k+1}$. So, at each iteration we replace the objective function $f(\vec{x})$ by its local quadratic approximation as follows:

$$f(\vec{\Delta x}) \approx \bar{f}(\vec{\Delta x}) = f(\vec{x}^k) + \vec{\nabla}^T f(\vec{x}^k)\vec{\Delta x} + \frac{1}{2}\vec{\Delta x}^T[H(\vec{x}^k)]\vec{\Delta x} \tag{5.47}$$

where $\vec{\Delta x} = \vec{x} - \vec{x}^k$. The constraint functions are approximated by their local linear approximations as follows:

$$h_i(\vec{\Delta x}) \approx \bar{h}_i(\vec{\Delta x}) = h_i(\vec{x}^k) + \vec{\nabla}^T h_i(\vec{x}^k)\vec{\Delta x}, i = 1 \cdots l. \tag{5.48}$$

$$g_j(\vec{\Delta x}) \approx \bar{g}_j(\vec{\Delta x}) = g_j(\vec{x}^k) + \vec{\nabla}^T g_j(\vec{x}^k)\vec{\Delta x}, j = 1 \cdots m. \tag{5.49}$$

Hence, at each iteration we solve the following quadratic programming subproblem:

$$\text{Minimize: } \bar{f}(\vec{\Delta x})$$
$$\text{subject to: } \bar{h}_i(\vec{x}) = 0, i = 1 \cdots l. \tag{5.50}$$
$$\bar{g}_j(\vec{x}) \leq 0, j = 1 \cdots m.$$

The output from solving this subproblem is $\vec{\Delta x}^k$ at the current iteration k; this $\vec{\Delta x}^k$ is used to update the current guess $\vec{x}^k$ using the update equation:

$$\vec{x}^{k+1} = \vec{x}^k + \vec{\Delta x}^k \tag{5.51}$$

The new iterate $\vec{x}^{k+1}$ is then used to compute new approximate functions $\bar{f}, \bar{h}_i, \bar{g}_j \ \forall i, j$ at $\vec{x}^{k+1}$. This process repeats until any of the stopping criteria is satisfied. The convergence stopping criterion is that all constraint functions of the original nonlinear programming problem are satisfied and the first order conditions of the objective function of the original nonlinear programming problem are satisfied at the current iterate. Another stopping criterion is when there is no significant change in $\vec{\Delta x}$ over subsequent iterations. Also the algorithm should stop when a maximum number of iterations is completed. **Algorithm** 5.2 shows the SQP algorithm.

One might think of the subsequent iterates $\vec{x}^k$ as a sequence that converges to a local minimum $\vec{x}^*$, of the given nonlinear programming problem, as k approaches ∞. One major advantage of the SQP method is that the iterates $\vec{x}^k$ need not be feasible; in other words the iterates $\vec{x}^k$ does not have to satisfy the equality or inequality constraints of the problem.

Algorithm 5.2 Sequential Quadratic Programming Algorithm

Select an initial guess for the solution $\vec{x}^0$

Select a stopping tolerance ε and a maximum number of iterations N_{itr}

k = 0

while k $< N_{itr}$ **do**

    Call the quadratic programming solver to optimize the optimization sub-problem, and output $\vec{\Delta x}^k$

    $\vec{x}^{k+1} = \vec{x}^k + \vec{\Delta x}^k$

    if $\vec{h} = \vec{0}$ and $\vec{g} \leq \vec{0}$ and KT conditions satisfied **then**

        STOP

    end if

    if $\| \vec{\Delta x} \| < \varepsilon$ **then**

        STOP

    end if

    if $k = N_{itr}$ **then**

        STOP

    end if

    k = k + 1

end while

VARIABLE-SIZE DESIGN SPACE OPTIMIZATION

Systems architecture optimization problems usually have categorical variables, and sometimes include multi-disciplinary optimization. In this type of problem, the size of the design space is sometimes variable. The focus in this part of the book is on the latter aspect of systems architecture optimization; the type of problems where the number of optimization variables is a variable.

Local optimization algorithms search for an optimal solution in the neighborhood of an initial guess. Global optimization methods search for optimal solutions in the whole design space. Genetic algorithms is one example of global optimization algorithms. The following chapters present recently developed variations on the genetic algorithm optimization method that make it more suitable for solving system architecture optimization problems.

Chapter 6

Hidden Genes Genetic Algorithms

6.1 Introduction to Global Optimization

Optimization problems may be classified based on the existence of constraints, the nature of the design variables, and the nature of the equations involved. Local optimization methods find a local minimum given an initial guess in its neighborhood. The methods that need the gradient information, such as those discussed in previous chapters, are local optimization methods. In reality, objective functions may have a large number of local minima. Finding the global minimum of a function is a more challenging problem, as compared to local minimum search. Consider for example, the interplanetary trajectory optimization problem described in **Section** 1.2.1. The cost of a trajectory is usually quantified in terms of the fuel needed to achieve the mission using this trajectory. The cost of a trajectory is usually a function of several variables. Assume we allow only two variables to change, e.g., the launch date and the flight time, and fix all other variables. In this case a typical cost function variation with these two variables is shown in **Fig.** 6.1. **Figure** 6.1 is a contour plot that maps a three-dimensional function on a two-dimensional plane using color coding. The color dictates the value of the cost function. As can be seen in **Fig.** 6.1, if we use a local optimization method and start with an initial guess for the time of flight near the value of 140 days, then the optimizer would find a local minimum that may not be a global optimal solution. There is a local minimum in the neighborhood of that initial guess that acts as an attractor for local optimization methods. It is of great impact if we have a tool that is able to search for the global optimal solution (identified in **Fig.** 6.1 as the "Optimal Solution") and get trapped in a local domain.

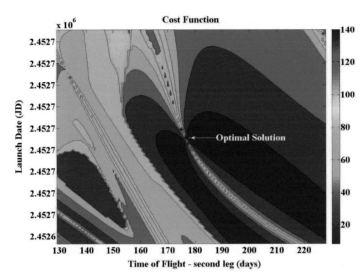

Figure 6.1: The trajectory optimization cost function has multiple local minima.

Several techniques were developed for global optimization of multi-modal functions. Some direct methods rely on the use of rectangular bounds and work by exploring the rectangular subregions. The whole region is sampled in the limit when the number of iterations is infinity. This method may suffer lack of efficiency in some problems. Global optimization can be achieved using local optimizers, if multiple start points are used. The global optimal solution is then picked from the local solutions. Clearly, several start points may lead to the same local solution, and hence render the algorithm inefficient. For a better computational efficiency, clustering methods replace each group of start points, that lead to the same local minimum by only one start point. The Branch and Bound method is effective in handling optimization problems with mixed design variables (integer and real). Yet, it requires the feasibility of dividing the design space, to create smaller sub problems (branching). The stochastic branch and bound method uses stochastic upper and lower estimates of the optimal value of the objective function in each sub problem.

There are several global optimization concepts; many of them are biologically inspired. By mid-twentieth century, Metaheuristic algorithms started to develop. Genetic algorithms (GAs), for example, were first introduced by John Holland in 1960s. In his early work, Holland introduced the crossover and reproduction operations. The crossover operation is key to produce solutions in new regions in the design space. For aerospace engineering applications, the evolutionary strategy search method was also developed with no crossover operation. Yet a mutation operation was introduced that is also a mechanism to generate offsprings in new regions. An elite solution was always stored in each iteration.

Inspired by the annealing process of metals, the simulated annealing (SA) search method was developed in 1980s. In simulated annealing, a new, randomly generated, solution replaces an initial guess solution under the control of a probabilistic criterion; and the process then repeats. The same decade also witnessed the development of Tabu search methods. Several bio inspired search methods were later invented. One method is the Particle Swarm Optimization (PSO) which is inspired by the swarming behaviour of birds. Each particle (individual) behaves based on both its own intelligence and the group intelligence. Another example is the Ant Colony Optimization (ACO) which is based on the intelligence of a colony of ants. Specifically, the ACO technique mimics the ant colonies cooperative behavior in finding the shortest path from a food source to their nest. In the ACO method, each layer represents a design variable. Each layer consists of nodes, that represent the permissible values of the corresponding design variable. In the ACO, each layer has a fixed number of nodes. To move from the nest to the food source, each ant in the colony moves through all layers and leaves a pheromone on the path. A global update rule is then used to update the globally best path. The ACO method has been implemented in several applications, of them are the traveling salesman problem, the job-shop scheduling problem, and the resource-constrained project scheduling problem. Differential Evolution (DE) is another optimization algorithm that proved to be more efficient than genetic algorithms in some applications. For scheduling optimization problems, the harmony search (HS) algorithm was proposed in 2001. There are also other optimization methods that proved to be efficient in some types of problems. For example the honey bee algorithm was found to be efficient in optimizing Internet hosting centers. An extension to that is the virtual bee algorithm and the artificial bee colony algorithm. Other interesting examples of bio inspired optimization algorithms include the firefly algorithm (FA) which was developed in 2007, cuckoo search (CS) algorithm (2009), the bat algorithm (2010), and the flower pollination algorithm (2012).

The methods presented in this book to handle VSDS optimization problems are based on global optimization methods. Specifically, the genetic algorithms is adopted to implement most of the concepts presented in this book. Hence a brief review for genetic algorithms and how they work is presented in the next section.

6.2   Genetic Algorithms

Standard GAs are search techniques based on the mechanics of natural selection and genetics [62]. The search for the optimal solution starts by initializing a population of solutions. This initialization is carried out by selecting random values for all the variables in each solution. This population evolves over time and converges to nearly-optimal solutions over a set of generations (iterations) of the algorithm. To set up a simple genetic algorithm, we start by coding (rep-

Figure 6.2: A chromosome (code) is a string of genes that represent a solution.

resenting) each solution in the population by a chromosome that consists of a string of genes. Each gene is the code of a variable, and hence the number of genes is the number of design variables. **Figure** 6.2 shows a typical chromosome that consists of N genes $g_1, g_2, \ldots, g_N$. The value of g_i determines the value of that variable in that solution. Associated with each chromosome (solution) is a fitness value which determines how fit is that solution. The fitness of the solution is determined based on the objective of optimization.

Starting with the initial population of solutions, GAs perform a number of operations on the current population to produce a new population of solutions; this new population of solutions is a new generation. At each generation, a Selection operator causes the highly fit solutions to survive in the population, and the less fit solutions to die. The GAs operations are then applied to the surviving solutions to create new solutions (offspring solutions) for the new generation. Of these GAs operations is the Crossover operation which is applied to randomly selected two individuals in the mating pool, bifurcate them at randomly selected sites, swap the string of genes, and create new individuals. The crossover operation is carried out with a probability p_c. A Mutation operation is also applied where the mutation operator goes through all the bits in all the genes of the population and modifies a particular bit with a mutation probability p_m. The same processes of selection, crossover and mutation are repeated in each generation to produce a newer generation until a stopping criteria is satisfied. As can be seen from the above brief, the GAs work on a fixed length chromosome, and hence the number of variables should be constant. GAs are then not suitable for the VSDS optimization problems.

GAs work with coded variables sets. Each variable is assigned a number of bits for coding. A *member* (or a *chromosome*) is the code for a stack of the binary codes for all variables together in one string. A group of chromosomes is called a *population*. Subsequent generations follow by performing a series of probabilistic operations on the current population. The basic operations that are used in all genetic algorithms are: reproduction, crossover, and mutation . Pairs of members are selected to undergo a crossover with a probability P_c, at a random point in the chromosome. Mutation is applied to all bits in all members in a generation with a probability P_m.

Given the above description for the different operations of genetic algorithms, there remains the question of why does it work? or in other words, how does it find the global optimal solution? There are different ways of explaining how the genetic algorithms work. One of the most simple ways to explain how it works is

The Schemata Theory, which is briefed below based on [62]. The Markov chain model of genetic algorithms is briefed in **Section** 6.2.2 based on [101] and [31].

6.2.1 Similarity Templates (schemata)

A schema (plural is schemata) is a subset of chromosomes with similarities at certain string positions. The order of a schema H, $o(H)$, is the number of fixed positions in the template (e.g., $o(011*1**) = 4$, where $*$ is the DO NOT CARE value.) The defining length of a schema, $\delta(H)$, is the distance between the first and the last specific positions in the string, e.g., $\delta(011*1**) = 4$. The number of members of the schema H at a given time t is $m = m(H,t)$. A particular schema S_i receives an expected number of members in the next generation under reproduction according to Eq. (6.1).

$$m(H,t+1) = \frac{m(H,t)f(H)}{f_p} \tag{6.1}$$

where $f(H)$ is a scalar function representing the average fitness for the strings of the schema H, f_p is the population average fitness. A schema survives a crossover operation with the probability P_s, where:

$$P_s = 1 - P_c \frac{\delta(H)}{(n_b - 1)} \tag{6.2}$$

and n_b is the number of bits in the member, P_c is the crossover probability. The survival probability of a schemata H can be approximated by: $1 - o(H)P_m$. So, by combining the reproduction, crossover and mutation operations, a particular schema receives an expected number of members in the next generation as given by:

$$m(H,t+1) = m(H,t)\left(\frac{f(H)}{f_p}\right)\left(1 - P_c\frac{\delta(H)}{(n_b - 1)} - o(H)P_m\right) \tag{6.3}$$

Equation (6.3) represents the Fundamental Theorem of GA.

6.2.2 Markov Chain Model

The stochastic dependency between successive populations is due to applying the selection, mutation, and crossover operators to the current population in producing the next population. Hence, the GA is a stochastic process in which the state of each population depends only on the state of the immediate predecessor population. Therefore, the GA can be modeled as a Markov process [31]. Several studies have investigated the convergence behavior of the GA explicitly using the Markov chain analysis [118, 86, 108, 31, 101]. The minimum conditions for convergence of GAs in the realm of Markov chain model can be found in details in [31, 18, 101]. Here, these conditions are briefly reviewed, and later utilized to derive the convergence conditions for the hidden genes genetic algorithms that is introduced later in this chapter. The GA is a Markov process and its transition matrix can be calculated. It will be shown that the GA transition matrix is reducible. Hence, the ergodic theorem for reducible transition matrix can be used to prove that ergodicity is a sufficient condition for convergence. It is assumed that this analysis is in the domain of binary genetic algorithms with bits as variables. The material in this section is a brief for the work in [101] and [31], presented here for completeness of the presentation. The Markov chain model of the hidden genes genetic algorithms presented later in this chapter is developed based on the same methodology briefed in this section. For more details of the Markov chain model of GAs, the reader is encouraged to review [101] and [31]. We start with a review for few basic definitions:

- Column-allowable matrix: a square matrix that has at least one positive entry in each column.

- Stochastic matrix: a non-negative matrix $\mathbf{A} = (a_{ij})_{i,j=1,...,n}$ is said to be stochastic if $\sum_{j=1,...,n} a_{ij} = 1$, for each $i = 1,...,n$.

- Arithmetic crossover: a crossover that linearly combines two parents to get one child. The child is the weighted average of the parents as follows:

$$C = \lambda P_{t_1} + (1 - \lambda) P_{t_2} \tag{6.4}$$

where C is the child, P_{t_1} and P_{t_2} are the parents, and λ is a random number in $(0, 0.5)$.

- Reducible matrix: if matrix $\mathbf{A} = (a_{ij})_{i,j=1,...,n}$ is non-negative and can be brought into the form $\begin{bmatrix} \mathbf{D} & \mathbf{0} \\ \mathbf{R} & \mathbf{T} \end{bmatrix}$ by applying the same permutations to rows and columns, it is called a reducible matrix. Note that $\mathbf{D}$ and $\mathbf{T}$ should be square matrices.

The finite state space S of a Markov chain has a size of $|S| = n$, where the states are numbered from 1 to n. Let l be the chromosome length, $M = 2^l$ be the constant population size, and $m = 2^{nl}$. Assume that the simple GA consists of three standard operations: selection (**S**), mutation (**M**), and crossover (**C**). To transform any state i to state j, the transition product matrix **CMS** is used and the convergence of the GA depends on this transition matrix [18]. The transition matrix of a finite Markov chain consists of the transition probabilities from state i to j, i.e., $\mathbf{P} = (p_{ij})$. For each entry, $\sum_{j=1}^{|S|} (p_{ij}) = 1$ for all $i \in S$. The GA transition product matrix (**CMS**) is a Markov probability matrix (**P**).

First few needed theorems and lemmata are listed here:

Lemma 1: Let **C**, **M** and **S** be stochastic matrices, where **M** is positive and **S** is column-allowable. Then the product **CMS** is positive [101].

Theorem 1: Let $\mathbf{P}$ be a primitive stochastic matrix. Then $\mathbf{P}^k$ converges as $k \to \infty$ to a positive stable stochastic matrix $\mathbf{P}^\infty = 1'\mathbf{p}^\infty$, where $\mathbf{p}^\infty = \mathbf{p}^0 . \lim_{k \to \infty} \mathbf{P}^k = \mathbf{p}^0 \mathbf{P}^\infty$ has nonzero entries and is unique regardless of the initial distribution [101].

Theorem 2: Let $\mathbf{P}$ be a reducible stochastic matrix defined as: $\begin{bmatrix} \mathbf{D} & 0 \\ \mathbf{R} & \mathbf{T} \end{bmatrix}$ where $\mathbf{D}$ is an $m \times m$ primitive stochastic matrix and $\mathbf{R}, \mathbf{T} \neq 0$. Then

$$\mathbf{P}^\infty = \lim_{k \to \infty} \mathbf{P}^k = \lim_{k \to \infty} \begin{bmatrix} \mathbf{D}^k & 0 \\ \sum_{i=0}^{k-1} \mathbf{T}^i \mathbf{R} \mathbf{D}^{k-i} & \mathbf{T}^k \end{bmatrix} = \begin{bmatrix} \mathbf{D}^\infty & 0 \\ \mathbf{R}_\infty & 0 \end{bmatrix} \quad (6.5)$$

is a stable stochastic matrix with $\mathbf{P}^\infty = l'\mathbf{p}^\infty$, where $\mathbf{p}^\infty = \mathbf{p}_0 \mathbf{P}^\infty$ is unique regardless of the initial distribution, and $\mathbf{p}^\infty$ satisfies: $p_i^\infty > 0$ for $1 \leq i \leq m$ and $p_i^\infty = 0$ for $m < i \leq n$ [101].

Theorem 3: The transition matrix of the GA with mutation probability $p_m \in (0,1)$, crossover probability $p_c \in [0,1]$ and proportional selection is primitive [101].

Corollary 1: The CGA with parameter ranges as in Theorem 1 is an ergodic Markov chain, i.e., there exists a unique limit distribution for the states of the chain with nonzero probability to be in any state at any time regardless of the initial distribution. This is an immediate consequence of Theorems 1 and 2 [101].

Theorem 4: The CGA with parameter ranges as in Theorem 3 does not converge to the global optimum [101].

Theorem 5: In an ergodic Markov chain the expected transition time between initial state i and any other state j is finite, regardless of the states i and j [101].

Theorem 6: The canonical GA as in Theorem 3 maintaining the best solution found over time after selection converges to the global optimum [101].

To maintain the best solution over time, the population is enlarged by adding the super individual to it. The term super individual is used for the solution that does not take part in the evolutionary process. Hence, the cardinality of the state space grows from 2^{nl} to $2^{(n+1)l}$. The super individual is placed at the leftmost position in the $(n+1)$-tuple and can be accessible by $\pi_0(i)$ from a population at state i, where $\pi_0(i)$ is a function that calls the super individual from population i.

The super individual does not take part in the evolutionary process, therefore, the extended transition matrices for crossover $\mathbf{C}^+$, mutation $\mathbf{M}^+$, and selection $\mathbf{S}^+$ can be written as [101]:

$$\mathbf{C}^+ = \begin{bmatrix} \mathbf{C} & & & \\ & \mathbf{C} & & \\ & & \cdots & \\ & & & \mathbf{C} \end{bmatrix}, \mathbf{M}^+ = \begin{bmatrix} \mathbf{M} & & & \\ & \mathbf{M} & & \\ & & \cdots & \\ & & & \mathbf{M} \end{bmatrix}, \mathbf{S}^+ = \begin{bmatrix} \mathbf{S} & & & \\ & \mathbf{S} & & \\ & & \cdots & \\ & & & \mathbf{S} \end{bmatrix}$$

$$(6.6)$$

Then we can write:

$$\mathbf{C}^+\mathbf{M}^+\mathbf{S}^+ = \begin{bmatrix} \mathbf{CMS} & & & \\ & \mathbf{CMS} & & \\ & & \ddots & \\ & & & \mathbf{CMS} \end{bmatrix} \tag{6.7}$$

where $\mathbf{C}^+$, $\mathbf{M}^+$, and $\mathbf{S}^+$ are block diagonal matrices and each of the 2^l square matrices $\mathbf{C}$, $\mathbf{M}$ and $\mathbf{S}$ are of size $2^{nl} \times 2^{nl}$, and $\mathbf{CMS} > 0$.

The upgrade matrix $\mathbf{U}$ is a matrix that upgrades the solutions in the population based on their objective function value (fitness). An intermediate state containing a solution with an objective value better than the super individual will upgrade to a state where the super individual equals the better solution. Let b be the best individual of the population at state i, excluding the super individual. By definition, $u_{ij} = 1$ if $f(\pi_0(i)) < b$, otherwise $u_{ii} = 1$. Therefore, there is one entry in each row and for every state j with $f(\pi_0(j)) < max[f(\pi_k(j))|k = 1\ldots n]$, the elements will be $u_{ij} = 0$ for all is. Hence, the structure of the upgrade matrix can be written as [101]:

$$\mathbf{U} = \begin{bmatrix} \mathbf{U}_{11} & & & \\ \mathbf{U}_{21} & \mathbf{U}_{22} & & \\ \cdots & \cdots & \cdots & \\ \mathbf{U}_{2^l,1} & \mathbf{U}_{2^l,2} & \cdots & \mathbf{U}_{2^l,2^l} \end{bmatrix} \tag{6.8}$$

where the sub-matrices $\mathbf{U}_{ab}$ are of size $2^{nl} \times 2^{nl}$. If the optimization problem has only one global solution, then only $\mathbf{U}_{11}$ is a unit matrix, and all other matrices $\mathbf{U}_{aa}$ with $a \geq 2$ are diagonal matrices with some zero diagonal elements, and some unit diagonal elements. Recall that in this Markov model for GA, $\mathbf{P} = \mathbf{CMS}$ and hence the transition matrix for the GA becomes:

$$\mathbf{P}^+ = \begin{bmatrix} \mathbf{P} & & & \\ & \mathbf{P} & & \\ & & \ddots & \\ & & & \mathbf{P} \end{bmatrix} \begin{bmatrix} \mathbf{U}_{11} & & & \\ \mathbf{U}_{21} & \mathbf{U}_{22} & & \\ \cdots & \cdots & \cdots & \\ \mathbf{U}_{2^l,1} & \mathbf{U}_{2^l,2} & \cdots & \mathbf{U}_{2^l,2^l} \end{bmatrix} = \begin{bmatrix} \mathbf{PU}_{11} & & & \\ \mathbf{PU}_{21} & \mathbf{PU}_{22} & & \\ \cdots & \cdots & \cdots & \\ \mathbf{PU}_{2^l,1} & \mathbf{PU}_{2^l,2} & \cdots & \mathbf{PU}_{2^l,2^l} \end{bmatrix} \tag{6.9}$$

Note that $\mathbf{PU}_{11} = \mathbf{P} > 0$. The sub-matrices $\mathbf{PU}_{a1}$, where $a \geq 2$, are gathered in a rectangular matrix $\mathbf{R} \neq 0$. Note that The $\mathbf{PU}_{1j} = 0$ where $\forall j > 1$. Then comparing Eq. (6.9) to Eq. (6.5), we can see that $\lim_{k\to\infty} \mathbf{P}^{+k}$ is unique regardless of the initial distribution, concluding in the convergence of the canonical GA.

Note that to make the extended transition matrix in the form of Eq. (6.9), we assumed that $\mathbf{C}$, $\mathbf{M}$, and $\mathbf{S}$ are stochastic, positive, and column-allowable. Therefore, the extended transition matrices $\mathbf{C}^+$, $\mathbf{M}^+$, and $\mathbf{S}^+$ are stochastic and positive. The above proof also shows that the $\mathbf{P}^+$ in Eq. (6.9) is a reducible matrix. Since $\mathbf{PU}_{11} > 0$ ($\mathbf{PU}_{11}$ corresponding to the $\mathbf{D}$ matrix in Theorem 2), then using Theorem 2 we can show that the GA converges to the optimal solution in the limit.

As discussed above, GAs recombine and mutate solutions with relatively high fitness to generate new solutions. In order to recombine different chromosomes together, their lengths must be equal, and hence the number of design variables must be the same in all solutions. In other words, the size of the design space must be fixed. Similarly, most other global optimization methods, such as simulated annealing and ant colony optimization, work on a fixed-size design space. In some applications, however, the size of the design space depends on the solution. Examples of this type of application include space trajectory optimization, system design optimization, pixel classification problems, and optimal grouping problems. These examples motivate the development of global optimization methods that can handle the problem of optimizing objective functions with Variable-Size Design Spaces (VSDSs). This chapter introduces the concept of Hidden Genes in genetic algorithms that enables GAs to handle VSDS optimization problems. This chapter is based on recent work published in [3, 49].

6.3 Fundamental Concepts of Hidden Genes Genetic Algorithms

The optimization of a simple VSDS objective function may be formulated as follows:

$$\text{minimize: } f(\vec{x}, J)$$
$$\text{subject to: } g_i(\vec{x}) \leq b_i \, , i = 1 \cdots m. \tag{6.10}$$
$$LB \leq \vec{x} \leq UB$$

where $\vec{x} = [x_1, x_2, ..., x_J]^T$, J is the number of design variables x_i, and UB and LB are the upper and lower bounds of the variables x_i, respectively. This section introduces the biologically inspired concept of hidden genes to address this type of VSDS optimization problems. The concept presented in this section was first published in reference [4].

6.3.1 The Hidden Genes Concept in Biology

In genetics, the deoxyribonucleic acid (DNA) is organized into long structures called chromosomes. Contained in the DNA are segments called genes. Each gene is an instruction for making a protein. These genes are written in a specific language. This language has only three-letter words, and the alphabet is only four letters. Hence, the total number of words is 64. The difference between any two persons is essentially because of the difference in the instructions written with these 64 words. Genes make proteins according to these words. Since, not all proteins are made in every cell, not every gene is read in every cell. For example,

Figure 6.3: Chemical tags (purple diamonds) and the "tails" of histone proteins (purple triangles) mark DNA to determine which genes will be transcribed (picture is modified from [111]).

an eye cell doesn't need any breathing genes on. And so they are shut off in the eye. Seeing genes are also shut off in the lungs. Another layer of coding tells what genes a cell should read and what genes should be hidden from the cell [106]. A gene that is being hidden, will not be transcribed in the cell. There are several ways to hide genes from the cell. One way is to cover up the start of a gene by chemical groups that get stuck to the DNA. In another way, a cell makes a protein that marks the genes to be read; **Fig.** 6.3 is an illustration for this concept. Some of the DNA in a cell is usually wrapped around nucleosomes but lots of DNA are not. The locations of the nucleosomes can control which genes get used in a cell and which are hidden [106].

6.3.2 Concept of Optimization using Hidden Genes Genetic Algorithms

In this chapter, the concept of hidden genes is implemented to hide some of the genes in a chromosome (solution); these hidden genes represent variables that should not appear in this candidate solution. In this section, the design variables are assumed to be coded in binary format. In a VSDS optimization problem, the number of design variables depends on the specific values of some of the design variables. Selecting different values for some of the design variables, changes the length of the chromosome. Different solutions have different number of design variables. So, in the design space, we have chromosomes of different lengths. Let L_{max} be the length of the longest possible chromosome. In the hidden genes concept, all chromosomes in the population are allocated a fixed length equal to L_{max}. In a general solution (a point in the design space), some of the variables

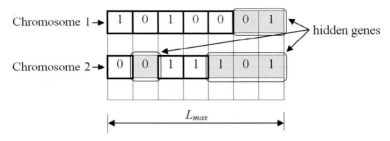

Figure 6.4: Hidden genes and effective genes in two different chromosomes.

(part of the chromosome) will be ineffective in objective function evaluations; the genes describing these variables are referred to as *hidden genes*. The hidden genes, however, will take part in the genetic operations in generating future generations. To illustrate this concept, consider **Fig.** 6.4. Suppose we have two chromosomes, the first chromosome is represented by five genes (represented by five binary bits in this example) and the second chromosome is represented by three genes. Suppose also that the maximum possible length for a chromosome in the population is fixed at 7. The first chromosome is augmented by two hidden genes and the second chromosome is augmented by four hidden genes. The hidden genes are not used to evaluate the the fitness of the chromosomes. Because all chromosomes have the same length, standard definitions of genetic algorithms operations can still be applied. Hidden genes will take part in all genetic operations. Mutation may alter the value of a hidden gene. A crossover operation will swap parts of the chromosome that may have hidden genes. A hidden gene in a parent may become an effective gene in the offspring. These hidden genes that become effective take part in the objective function evaluations in the new generations. **Figure** 6.5 shows a simple example for two parents with hidden genes and the resulting children after a crossover operation. In **Fig.** 6.5, the digit number is listed on the top of the figure. The digit number 6 determines the start location of the hidden genes, whereas the digits number 4 and 5 determine the number of hidden genes in the chromosome. The crossover point is between digits 4 and 5. As can be seen from **Fig.** 6.5, the parent chromosomes swap the digits 0–4. As a result of this swap: the gene values change in both of the resulting chromosomes, the number of hidden genes and/or the location of the hidden gene change.

6.3.3   *Outline of a Simple HGGA*

A simple version of the HGGA implements the standard GA crossover operation. The differences between the HGGA in this case and a standard GA are in how the chromosome length is computed and in how the fitness function is evaluated. The chromosome length is fixed among all solutions in the population and its length is selected to represent the solution that has the maximum number of

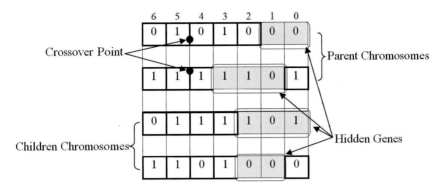

Figure 6.5: Crossover operation in HGGA.

design variables. In evaluating the fitness of a solution, the chromosome hidden genes are excluded. **Algorithm** 6.1 shows an outline for the HGGA.

Algorithm 6.1 The hidden genes genetic algorithm

Determine the maximum number of design variables DV_{max}
Compute the chromosome length C_L: $C_L = DV_{max}$
Create a population of chromosomes
Initialize the population
Evaluate initial population

■ Determine the hidden genes in each chromosome

■ Evaluate fitness of each chromosome - Exclude hidden genes in evaluating fitness

while (some convergence criteria is not satisfied) **do**
    Perform competitive selection
    Apply genetic operators to generate new chromosomes
    Evaluate solutions in the population
end while

Example 6.1. Branin Function Optimization Using Simple HGGA

In this section, an experimental evaluation of the proposed approach on the benchmark Branin function is conducted. The Branin function $f(y_1, y_2)$ is defined as:

$$f(y_1, y_2) = \left(y_2 - \frac{5.1}{4\pi^2} y_1^2 + 5\frac{y_1}{\pi} - 6 \right)^2 + 10 \left(1 - \frac{1}{8\pi} \right) cos(y_1) + 10 \quad (6.11)$$

$$y_1 \in [-5, 10] \text{ and } y_2 \in [0, 15]$$

The Branin function has three global minima. First, the standard GA is implemented. The chromosome has two genes, one for each variable y_1 and y_2. The resulting solution has $f = 0.3979$, $y_1 = 3.1416$, and $y_2 = 2.2749$. This GA implementation has a population of 100 members and the number of generations is 100.

To test the HGGA, two cases are considered. The first case is implementing the HGGA to the same problem. The chromosome, in the HGGA implementation, consists of three genes: y_1, y_2, and n, where n is an integer that specifies which gene of y_1 and y_2 is hidden. If $n = 1$, then the gene representing y_2 is hidden. If $n = 2$, then the gene representing y_1 is hidden. If $n = 0$, then none of the genes is hidden. The result of implementing the HGGA is $f = 0.3979$, $y_1 = 3.1416$, $y_2 = 2.2750$, $n = 0$. The population size is 100 and the number of generations is 100. The solution obtained from the HGGA is the same as that obtained from the GA, and the chromosome of the obtained solution has no hidden genes. The second case tests the new capability of the HGGA. Consider another function $f' = f(y_1, y_2) + y_3^2$, where $y_3 \in [0\,10]$. Suppose that the number of variables in f' is variable: two or three. Clearly, the minimum value for f' is the same as that of f. Hence, one global minimum of f' has only two variables y_1 and y_2 or three variables with $y_3 = 0$. The HGGA is implemented in this problem with a chromosome of four genes: y_1, y_2, y_3, and n, where $n \in \{0, 1, 2, 3, 4, 5, 6\}$ and specifies which gene(s) are hidden, as shown in **Fig.** 6.6 where the hidden genes are marked by 'H'. The result of running the HGGA is $f' = 0.3979$, $y_1 = 3.1416$, $y_2 = 2.2750$, $y_3 = 0.0001$, $n = 0$. Clearly the obtained solution corresponds to the case described above where $f' = f$ and y_3 vanishes. To force the HGGA to search for solutions that have hidden gene(s), the range of n is reduced to be $n \in \{1, 2, 3, 4, 5, 6\}$. Hence, any solution will have at least one hidden gene. Running the HGGA for this new case results in the solution: $f' = 0.3979$, $y_1 = 3.1416$, $y_2 = 2.2750$, $y_3 = 0.8387$, $n = 3$. This solution has y_3 as a hidden gene and hence its value does not affect the cost f'. The optimal values of y_1 and y_2 are those minimizing f.

n	0	1	2	3	4	5	6
y_1			H		H	H	
y_2		H			H		H
y_3				H		H	H

Figure 6.6: Assigned hiddene genes for different value of n.

6.4 The Schema Theorem and the Simple HGGA

This section presents an explanation for how the HGGA works within the realm of Holland Schema Theorem. The Schema theorem (The Fundamental Theorem of GAs) examines each of the fundamental operations in GAs (reproduction, crossover, and mutation), and provides a lower bound on the expectation for the schema population, for each one of these operations. The following are details for the effects of hidden genes on each of these operations.

6.4.1 Reproduction

To illustrate the effect of hidden genes on the reproduction operation, the following simple example is considered. Assume it is desired to maintain a certain temperature T_o at point 'O' in the rectangular space shown in **Fig.** 6.7. The temperature at point 'O' will be controlled by the air supply to the space. The number of air inlets and the amount of air per inlet are design variables to be optimized. The objective of optimization is to minimize the total amount of inlet air into the space. To make the presentation simple, assume that the maximum possible number of inlets is four, and the positions of all inlets are fixed, as shown in **Fig.** 6.7. Let the range of air flow rate variable for each of the inlets be 0–3 volume flow rate units, and the desired temperature at point 'O' is $T_o = 1.5$ temperature units. The design space includes solutions with one to four air inlets. Suppose that the temperature at point 'O' is related to the design variables by Eq. 6.12.

$$T_o = \frac{1}{n}\Sigma_{i=1}^{n}\frac{q_i}{r_i} \tag{6.12}$$

Equation 6.12 can be used to compute one of the air flow rates as follows:

$$q_n = r_n \times \left(nT_o - \Sigma_{i=1}^{n-1}\frac{q_i}{r_i}\right) \tag{6.13}$$

So, we have three independent design variables for the air flow rates. We select a chromosome that includes all possible design variables; some of the genes will be hidden, though, in some of the chromosomes in a population. **Figure** 6.8 shows typical chromosomes of this problem. The optimal solution for this problem is to have two inlets, at points '1' and '2'. The inlet at point '2' should have the

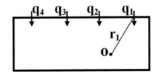

Figure 6.7: Temprature control system.

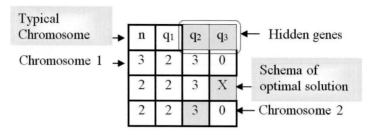

Figure 6.8: Chromosomes of the temperature control example.

maximum air flow rate (3 units), the air flow rate for the inlet at point '1' will have air flow rate of 2 units, and no air inlets are needed at points '3' and '4'.

The design variable n, in **Fig.** 6.8, is an integer variable that specifies the number of effective variables q_i. Maximum value of n is 3. The variable q_4 is always computed using Eq. (6.13). The mechanism for specifying the locations of the hidden genes is omitted in this discussion, without loss of generality, to make the presentation simpler. In **Fig.** 6.8, chromosome 1 represents the optimal solution, where q_2 takes the maximum value. Chromosome 2 has q_2 as hidden. Chromosome 2, which corresponds to a solution with no inlet at point '2', has poor fitness, and hence will have low probability of being selected for the next generation. A chromosome with effective q_2 and a hidden q_3 will have higher fitness, and hence higher probability of being selected for the next generation. It is noted here that having a hidden gene is not equivalent to having it effective with a value of zero despite the fact the chromosome fitness is the same in both cases. Each of these two genes has different effects on the subsequent generation. Also, a chromosome with a hidden gene belongs to schema that are different from the schema to which the chromosome belongs to if the gene is effective with a value of zero.

The objective function to be optimized is:

$$J = \sum_{i=1}^{4} q_i \tag{6.14}$$

where some/all of the variables, q_i, are represented by hidden genes. We can construct another function, G, that is equivalent to J, but with all of its variables represented by effective genes (all chromosomes have equal fixed length):

$$G = \sum_{i=1}^{4} p_i \tag{6.15}$$

where $p_i = \bar{p}_i q_i$, $\bar{p}_i = 0$ if q_i is hidden, and $\bar{p}_i = 1$ otherwise. Assume the string $A = nq_1q_2q_3$ represents a solution for the optimization of the objective function

n	q_1	q_2	q_3	J		p_1	p_2	p_3	G
1	3	1	0	3	→	3	0	0	3
3	0	1	3	4	→	0	1	3	4
2	0	2	1	1	→	0	0	1	1
1	3	2	0	2	→	0	2	0	2
1	1	0	2	2	→	0	0	2	2
2	3	2	0	5	→	3	2	0	5

Figure 6.9: Typical population for the temperature control problem.

J and let $\mathbf{A}(t)$ be the population at generation t. Assume the string $B = p_1 p_2 p_3$ represents a solution for the optimization of the objective function G and let $\mathbf{B}(t)$ be the population at generation t. **Figure** 6.9 shows a typical population of 6 members, the fitness of each member J, the corresponding p chromosomes, and the corresponding fitness G of each chromosome p. The hidden genes are shaded.

A schema, S_A, in the **A** population has a corresponding schema, S_B, in the **B** population, where both schemata represent the same set of solutions and have the same average fitness. This paragraph details why S_A and S_B have the same average fitness. The gene p_i takes the same value as q_i if the latter is effective. If q_i is hidden, then $p_i = 0$. **Figure** 6.10 shows three examples of schemata in **A** and their corresponding schemata in **B**. The shaded genes in **Fig.** 6.10 are hidden genes. In each row the schema S_A (left) corresponds to the schema S_B (right). In the first row, $S_{A1} = 1 * HH$ where q_1 is effective and takes the DO NOT CARE value $*$, q_2 and q_3 are hidden (H). The schema S_{A1} describes a subset with four members $\{10HH, 11HH, 12HH, 13HH\}$. The corresponding schema, $S_{B1} = *00$, has three effective genes. Since q_2 and q_3 are hidden, p_2 and p_3 take the value 0. The schema S_{B1} describes a subset with four members $\{000, 100, 200, 300\}$. Using Eqs. (6.14) and (6.15), it can be shown that the fitness G of the solution

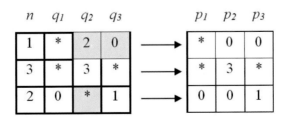

n	q_1	q_2	q_3		p_1	p_2	p_3
1	*	2	0	→	*	0	0
3	*	3	*	→	*	3	*
2	0	*	1	→	0	0	1

Figure 6.10: Corresponding schemata in the populations $\mathbf{A}(t)$(left) and $\mathbf{B}(t)$(right).

$p_1 p_2 p_3 = 100$ is the same as the fitness J of the solution $nq_1 = 11HH$. Similarly, each of the members of S_{A1} has a cost (objective function value) that is equal to a corresponding member in S_{B1}. Hence the average fitness of the schema S_{A1} is equal to the average fitness of the schema S_{B1}. In a similar way, it is possible to show that the average fitness of the schema S_{A2} is equal to the average fitness of the schema S_{B2}, and the average fitness of the schema S_{A3} is equal to the average fitness of the schema S_{B3}, in the second and third rows of **Fig.** 6.10, respectively.

From Eq. (6.1), during reproduction, a particular schema S_B will receive an expected number of members in the next generation according to the equation:

$$m(S_B, t+1) = \frac{m(S_B, t)G(S_B)}{G_p(t)} \tag{6.16}$$

Where $G(S_B)$ is the average fitness of the members represented by the schema S_B, and $G_p(t)$ is the average fitness of the **B**(t) population. Suppose we start with the initial random population **A**(0). We generate an initial population **B**(0), by computing the corresponding member in **B**(0) for each member in **A**(0). As can be seen in **Fig.** 6.9, the average fitness is equal in both populations, i.e., $G_p(0) = J_p(0)$. As we have found above, for every schema S_A, there is a corresponding equivalent schema S_B that has the same average fitness. Hence, $G(S_B) = J(S_A)$. We also found that the subset of members in each of S_A and S_B is the same: $m(S_B, 0) = m(S_A, 0)$. Consequently, we can conclude that:

$$m(S_A, 1) := \frac{m(S_A, 0)J(S_A)}{J_p(0)} = \frac{m(S_B, 0)G(S_B)}{G_p(0)} := m(S_B, 1) \tag{6.17}$$

Hence, using Eq. (6.1) it can be concluded that in subsequent generations, a schema S_A will receive an expected number of members the same way the corresponding schema S_B receives. As an example, consider the schema S_{A1} described above. The number of member is S_{A1} is $m(S_{A1}, 0) = 4$. The cost of each of the four members is computed using Eq. (6.14) and are listed in **Fig.** 6.11. The av-

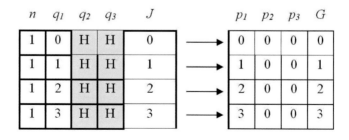

Figure 6.11: Memebers of S_{A1} and S_{B1} and their costs.

erage cost of all members in the schema S_{A1} is $J(S_{A1}) = \frac{6}{4}$. The population cost (see **Fig.** 6.9) is $J_p(0) = \frac{17}{6}$. Hence, the number of members in the schema S_{A1} in the next generations is:

$$m(S_A, 1) = \frac{m(S_A, 0)J(S_A)}{J_p(0)} = \frac{36}{17} \tag{6.18}$$

Similarly, the number of member is S_{B1} is $m(S_{B1}, 0) = 4$. The cost of each of the four members is computed using Eq. (6.15) and are listed in **Fig.** 6.11. The average cost of all members in the schema S_{B1} is $G(S_{B1}) = \frac{6}{4}$. The population cost (see **Fig.** 6.9) is $G_p(0) = \frac{17}{6}$. Hence, the number of members in the schema S_{B1} in the next generations is:

$$m(S_B, 1) = \frac{m(S_B, 0)G(S_B)}{G_p(0)} = \frac{36}{17} \tag{6.19}$$

6.4.2 Crossover

In order to explain how the hidden genes affect the increase (decrease) of the members in a schema, a modified defining length is introduced. The modified defining length of a schema, $\delta_h(S_A)$, is the distance between the first non-hidden and the last non-hidden specific positions in the string, e.g., $\delta_h(01\mathit{1} * 1\mathit{1} *) = 4$, where the italic genes are hidden. In standard genetic algorithms, a shorter schema has higher probability of surviving a crossover operation. The defining length of a schema can only get shorter if we make some of the genes hidden in a schema, so we can write:

$$\delta_h(S_A) \le \delta(S_B) \tag{6.20}$$

Recall Eq. (6.2), the probability of survival of a schema with hidden genes in a crossover operation, P_{sh}, will be:

$$P_{sh} = 1 - P_c \frac{\delta_h(S_A)}{(n_b - 1)} \ge 1 - P_c \frac{\delta(S_B)}{(n_b - 1)} \tag{6.21}$$

The standard genetic algorithm survival probability of a schema in the absence of hidden genes is, then, a conservative lower bound on the survival probability of that schema if some of its genes are hidden.

6.4.3 Mutation

If a hidden gene is mutated, there will be no effect in terms of member fitness. The modified order of a schema with a hidden gene, $o_h(S_A)$, is the number of non-hidden fixed positions in the schema (e.g., $o(01\mathit{1} * 1 * \mathit{0}) = 3$.) One or more hidden genes reduces the order of a schema. The survival probability, then, can be approximated as:

$$P_{sh} = 1 - o_h(S_A)P_m \ge 1 - o(S_B)P_m \tag{6.22}$$

If a hidden gene is mutated in a chromosome, the fitness of this chromosome does not change due this mutation. However, offsprings from this chromosome may be affected by this mutation.

From the preceding discussion, it can be concluded that the standard genetic algorithm expectation for the increase (decrease) in the number of members of a schema due to the combined effect of reproduction, crossover, and mutation is a conservative lower bound in the case of using hidden genes. This can be written as:

$$
\begin{aligned}
m(S_A, t+1) &= m(S_A, t)\left(\frac{J(S_A)}{J_p}\right)\left(1 - P_c\frac{\delta_h(S_A)}{(n_b - 1)} - o_h(S_A)P_m\right) \\
&\geq m(S_B, t)\left(\frac{G(S_B)}{G_p}\right)\left(1 - P_c\frac{\delta(S_B)}{(n_b - 1)} - o(S_B)P_m\right) \quad (6.23)
\end{aligned}
$$

6.5   Hidden Genes Assignment Methods

This section addresses the question of which genes are selected to be hidden in the HGGA, in each chromosome, in each generation, during the search for the optimal solution (optimal configuration). Several concepts for evolving the hidden genes are presented in this section. These concepts were first published in [5]. This mechanism of assigning hidden genes in a chromosome is vital for the efficient performance of HGGA. In the simple HGGA described in **Section** 6.3, the mechanism that was used to assign the hidden genes (the feasibility mechanism) was primitive. The feasibility mechanism rule assumes initially no hidden genes in a chromosome; if the obtained chromosome is feasible then there is no hidden genes. If the solution is not feasible, then starting from one end of the chromosome the algorithm hides genes - one by one - until the chromosome becomes a feasible chromosome.

In genetics, as discussed in **Section** 6.3.1, a cell makes a protein that marks the genes to be read. Inspired by genetics, it is proposed to use a tag for each of the genes that have the potential to be hidden (configuration gene) [4]. This tag determines whether that gene is hidden or not. The tag is implemented as a binary digit that can take a value of '1' or '0', as shown in **Fig.** 6.12. For each gene x_i that can be hidden, a *tag_i* is assigned to decide whether it is hidden or not. If *tag_i* is 1, then x_i is hidden, and if it is 0, x_i is active.

Figure 6.12: HGGA and the tags concept.

The values of these tags evolve dynamically as chromosomes change during the optimization process. Preliminary work in [4] suggests mechanisms for tags evolution. This section presents two different concepts for tags evolution. Both concepts are shown below with different variations on each concept.

6.5.1   *Logical Evolution of Tags*

During the chromosome crossover operation, the tags for the children are computed from the tags of the parents, using the logical OR operation. Using a logical operation is not new to GA. Reference [92] used them in the chromosomes crossover operation in GAs. Reference [74] presents a crossover operation where the similar genes in the parents are copied to the two children while the remaining genes in each child are randomly chosen from the two parents. Here, however, the logical operator is applied only on the tags. The crossover of two parents results in two children. Three logics are here presented for the logical evolution of tags:

Logic A: For one child, a gene is *hidden* if the same gene is hidden in any of the parents (Hidden-OR). For the second child, a gene is *active* if the same gene is active in any of the parents (Active-OR). It is possible to think of the logic of the second child as the AND logical operation when used with the hidden state – that is in the second child a gene is *hidden* if the same gene is hidden in both of the parents (Hidden-AND). The resulting children from crossover operation is shown in **Fig.** 6.13 along with the resulting new tags for the children.

Logic B: The Hidden-OR logic is used for both children. Even though the tags will be the same in both children, the two children represent two different solutions because they have different gene values.

Logic C: The Active-OR logic is used for both children.

6.5.2   *Stochastic Evolution of Tags*

In this concept, the tags are evolved using crossover and/or mutation operations, in a similar way to that of the design variables. Eight mechanisms are investigated using this concept. These mechanisms are:

Mechanism A: tags evolve through a mutation operation with a certain mutation probability. In this mechanism, the tags are separate from the design variables in the chromosome.

Mechanism B: tags evolve through mutation and crossover operations. In this mechanism, the tags are considered as discrete variables similar to the design variables in the chromosome. The tags are appended to the design

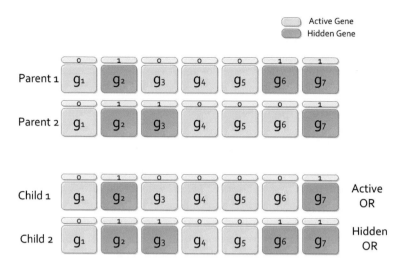

Figure 6.13: Tags of children as computed using the logic A.

variables; and hence their values are optimized along with the other variables through the selection, crossover, and mutation operations. In this mechanism the number of design variables is increased. The computational cost of evaluating the cost function is not changed though.

Mechanism C: tags evolve through a crossover operation. In this mechanism, the tags are considered as discrete variables similar to the design variables in the chromosome; yet only crossover operation can be applied to the tags.

Mechanism D: tags evolve through a mutation operation. In this mechanism, the tags are considered as discrete variables similar to the design variables in the chromosome; yet only mutation operation is applied to the tags. The mutation probability in this mechanism is the same as that used to mutate the main chromosomes.

Mechanism E: tags crossover independently from the genes. In other words, the tags may swap while the genes do not, or vise versa. This mechanism can be interpreted as a 2-*D* multiple crossover operator, one direction through tags and one direction through genes. Before applying the crossover operator, tags undergo a mutation operation.

Mechanism F: tags crossover using a logical fitness guided (arithmetic) crossover in which two intermediate chromosomes are produced. In these intermediate chromosomes, the genes are produced from a single crossover operator on parents and the tags are the outcome of the Active-OR logic on the parents' tags. In other words, parent X will have intermediate offspring Xx, and parent Y will have intermediate offspring Yy using the arithmetic crossover operation. The actual offspring is then created by a fitness guided crossover operation on the parents, and it is closer to the parent whose intermediate offspring has better cost.

Mechanism G: the arithmetic crossover is used with a modified cost function based on the number of genes that are hidden. The offspring is biased toward better parent (lower cost) with more hidden genes. The cost function is modified as follows: $f_{modified}(X) = f(X) - \sum_{i=1}^{M}(flag_i)$, where $flag_i$ is the value of the tag for gene i.

Mechanism H: the arithmetic crossover is used with a modified cost function based on the number of hidden genes. The offspring is biased toward better parent (lower cost) with less hidden genes. The cost function is $f_{modified}(X) = f(X) + \sum_{i=1}^{M}(flag_i)$.

The HGGA method presented in this chapter is relatively simple to implement. The genes undergo the standard GAs operations while the binary tags undergo different operations depending on the selected mechanism/logic as detailed above. Hence any existing code for GAs can be appended by a code that handles the tags, to create the proposed VSDS GAs.

6.6 Examples: VSDS Mathematical Functions

Multi-minima mathematical functions can be very useful in testing new optimization algorithms. For example, some mathematical functions can be used to tune the tags mutation probability for the different mechanisms presented in this paper, before they are used to solve more computationally intense problems. Three benchmark mathematical optimization problems were modified to make them VSDS functions, so that they can be used to test the HGGA. These functions are: the Egg Holder, the Schwefel 2.26, and the Styblinski-Tang functions. These functions have special structure; they are designed to test the effectiveness of global optimization algorithms. The general concept of modifying these functions to be VSDS is as follows. Consider the optimization cost function defined as:

$$F(X) = \sum_{i=1}^{N} f(x_i) \tag{6.24}$$

For each gene x_i, if its associated tag, tag_i, is 1 (hidden) then the gene x_i is hidden and hence $f(x_i)$ is set to zero (does not exist). This is consistent with the physical system test cases presented in **Section** 2.2. Unlike the hidden gene tags, the chromosomes evolve through the standard GA selection, mutation and crossover operations. For the chromosomes, a single point crossover and an adaptive feasible mutation operators are selected.

In the test cases presented in this section, the population size is 400, the number of generations is 50, the elite count is 20, the genes mutation probability is 0.01, and the crossover probability is 0.95. For the purpose of statistical analysis, each numerical experiment is repeated n times (n identical runs) and the success rate in finding the optimal solution is assessed. The success rate S_r is computed as $S_r = j_s/n$, where n is the number of runs and j_s is a counter that counts how many times the optimal solution is obtained as the solution at the end of an experiment [114].

6.6.1   *Examples using Stochastically Evolving Tags*

Example 6.2. The Schwefel 2.26 Function
The Schwefel 2.26 function is defined as follows:

$$f(X) = \sum_{i=1}^{N} f_i(x_i) = \sum_{i=1}^{N} -\frac{1}{N} x_i sin(\sqrt{x_i}). \tag{6.25}$$

subject to $-500 \leq x_i \leq 500$.
Here it is assumed that $N = 5$ (maximum possible number of variables is 5). Minimizing $f(X)$, the optimal solution is known, and it is $f_{min} = -418.9829$. First, the standard GA is used to solve this problem. In using the standard GA, all variables are active. The results show that a minimum of -418.8912161 is obtained using the standard GA. For the HGGA, there are N tags in this case. The results of HGGA Mechanism A for different mutation probability values are presented in **Table** 6.1.
As shown in **Table** 6.1, the minimum obtained function value is -418.8967549 which is lower than the solution obtained using the standard GA.

Table 6.1: Mechanism A: Tags separated and mutated.

Mutation Probability of Tags	Function Value	Occurrence Probability
0.00001	−418.8803137	100
0.0001	−418.889191	100
0.001	−418.8967549	100
0.005	−418.1629931	80
0.01	−418.2677072	80
0.02	−416.8354092	20
0.03	−418.8471854	90
0.04	−417.941699	90
0.1	−418.7225599	50

The occurrence probability (success rate) for Mechanism A is 100%. For Mechanism B, the minimum obtained function value is -418.4088183 with occurrence probability of 100%. For Mechanism D, the results show that the minimum obtained function value is -418.1721324 with occurrence probability of 90%. For Mechanism C, the minimum obtained function value is -418.2119 with occurrence probability of 100%. The results of these four HGGA mechanisms on the Schwefel 2.26 function are summarized in **Fig.** 6.14, where the percentages on the figure indicate the success rate of the mechanism.

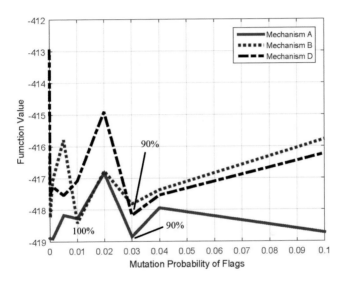

Figure 6.14: Comparison of Three HGGA tags mechanisms - using Schwefel 2.26 Function.

Example 6.3. The Egg Holder Function
The Egg Holder function is defined as follows:

$$f(X) = \sum_{i=1}^{N-1} f_i(x_i, x_{i+1})$$

$$= \sum_{i=1}^{N-1} -(x_{i+1} + 47)sin(\sqrt{x_{i+1} + 0.5x_i + 47})$$

$$-x_i sin(\sqrt{x_i - x_{i+1} - 47}) \qquad (6.26)$$

where $-512 \leq x_i \leq 512$.

This is an interesting case study for the HGGA. Each function $f_i(x_i, x_{i+1})$ is a function of two variables. Hence if a variable x_i is hidden, this does not necessarily mean the function f_i goes to zero. There are few ways of handling this type of functions depending on what these functions represent in a physical system. For example, in the interplanetary trajectory optimization problem (will be discussed in detail later in this paper), an event of planetary fly-by is associated with few variables, of them are the fly-by height and the fly-by plane orientation angles. If one of these variables is hidden then that implies the whole fly-by event is hidden and hence the other variables in this event are also hidden. This example suggests that a function $f_i(x_i, x_{i+1})$ (could be representing the cost of a fly-by) would assume a value of zero if any of the variables x_i or x_{i+1} is hidden. In other situations, however, this is not the case. A function $f_i(x_i, x_{i+1})$ may have a non-zero value despite one of the variables x_i or x_{i+1} being hidden. Here in this mathematical function, a tag tag_i is assigned to each function f_i and hence the tag_i value determines whether the value of the function f_i is zero or not.

First, the standard GA is used to optimize this function assuming $N = 5$. The best solution found by the standard GA has a function value of -3657.862773. For the HGGA, there are $N-1$ tags for the $N-1$ functions f_i. Different mutation probability values are tested for Mechanism A and the results are presented in **Table** 6.2 for $N = 5$.

As shown in **Table** 6.2, the mutation probability of 0.03 has the lowest cost function of -3692.314081 with occurrence probability of 60%. This solution is better than the solution obtained using the standard GA.

A similar analysis is conducted for Mechanism B. The results show that the lowest obtained function value is -3599.845154 with mutation probability of 0.0001 and occurrence probability of 10%. For Mechanism D, the crossover and mutation operations are applied to the N design variables (x_i) and only mutation is applied for the rest $N-1$ variables (the tags). The results show that the lowest obtained function value is -3484.255119 with mutation probability of 0.01 and occurrence probability of 10%. For Mechanism C, the tags are variables (to-

Table 6.2: Mechanism A: Tags separated and only mutated.

Mutation Probability of Tags	Cost Value	Occurrence Probability
0.00001	−3615.942419	40
0.0001	−3216.79243	100
0.001	−3390.917369	10
0.005	−3655.26687	20
0.01	−3499.743532	10
0.02	−3424.28003	10
0.03	−3692.314081	60
0.04	−3640.861368	20
0.1	−3587.802092	10

tal number of GA variables $2N - 1$) and crossover and mutation operations are carried out on the N design variables. Only the crossover operation is applied on the next $N - 1$ variables (the tags). The minimum obtained function value is -3559.5751 with occurrence probability of 10%. Mechanism A produced the best solution in this test for the Egg Holder function. The results of these tests are summarized in **Fig.** 6.15, where the percentages on the figure indicate the success rate of the mechanism.

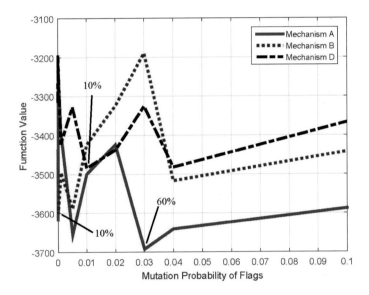

Figure 6.15: Comparison of Three HGGA tags mechanisms - using Egg Holder Function.

Example 6.4. The Styblinski-Tang Function
The Styblinski-Tang function is defined as:

$$f(X) = \sum_{i=1}^{N-1} f_i(x_i, x_{i+1}) = \sum_{i=1}^{N} \frac{1}{2}\left(x_i - 16x_i^2 + 5x_i\right). \tag{6.27}$$

subject to $-10 \leq x_i \leq 10$. It is assumed that $N = 5$. Standard GA results in a minimum obtained function value of -195.8304861. For the HGGA Mechanism A, the obtained function value is -195.8306992, with a mutation probability of 0.03, with occurrence probability of 100%.

For Mechanism B, the results show that the minimum obtained function value is -195.828109 with mutation probability of 0.01 and occurrence probability of 100%. For Mechanism D, the results show that the minimum obtained function value is -195.8261996 with mutation probability of 0.005 and occurrence probability of 100%. For Mechanism C, the minimum obtained function value is -195.82982 with occurrence probability of 100%.

Example 6.5. Mechanisms E, F, G, and H
The GA parameters used in this example are listed in **Table** 6.3.

Table 6.3: Genetic algorithm options in MATLAB®.

Option	Value
Population Size	400
Number of Generation	50
Mutation Probability	0.01
Elite Count	20
Crossover Fraction	0.95
TolFun	$1e-6$
TolCon	$1e-6$

In general, standard GA are not suitable for solving VSDS problems. However, a significant advantage of using the above modified mathematical functions is the possibility of using standard GA if we assume all variable are active (not hidden). If the optimal solution has x_j hidden $\forall j \in \Gamma$, and $\Gamma \subseteq \{1, 2, \cdots, N\}$, then the standard GA can find that optimal solution, if we assume all variables are not hidden. In such case, the optimal solution that the standard GA would search for is x_j^* where $f(x_j^*) = 0$, $\forall j \in \Gamma$.

The mechanisms E, F, G, and H are tested on four selected mathematical optimization functions. These functions are: the Egg Holder, the Schwefel 2.26, the Styblinski-Tang, and the Ackley 4 functions. Each test case is simulated 20 times, with population size of 400 and 50 number of generations. In Eq. (6.27), if f_i is a function of x_i only, there are N tags, and if f_i is a function of x_i and x_{i+1}, then there are $N-1$ tags. In all the problems, the number of variables without tags is 5. Based on the simulation results (first four rows in **Table** 6.4), Mechanism E is the best choice among four proposed mechanisms for 2 out of 4 test cases. Moreover, comparing the four proposed mechanisms, Mechanisms G and H result in highest cost function with lowest occurrence probability in all the cases.

Table 6.4: Objective function values of all the test cases.

	Mechanism E	Mechanism F	Mechanism G	Mechanism H	Simple HGGA
Egg Holder function	−3552.8947, 50%	−3644.2279, 50%	−2571.8028, 10%	−2134.7217, 10%	−2749.2646, 30%
Schwefel 2.26 function	−418.2850, 100%	−418.0799, 100%	−375.3514, 5%	−303.5320, 10%	−335.1844, 100%
Styblinski-Tang function	−195.8298, 100%	−195.8278, 100%	−191.8393, 10%	−194.5325, 10%	−156.6646, 100%
Ackley 4 function	−12.2281, 10%	−12.4931, 20%	−11.6045, 10%	−11.8215, 10%	−7.8898, 95%

6.6.2 Examples using Logically Evolving Tags

The logical mechanisms for tags assignment presented in **Section** 6.5.1 are tested on three of the mathematical functions described above. The results are summarized in **Table** 6.5 that lists the obtained solution using each Logic, for each of the functions.

Table 6.5: Solutions obtained using logical evolution methods of HGGA.

Function	Logic A	Logic B	Logic C
Egg Holder	−3386.2499	−3192.6139	−3634.1438
Schwefel 2.26	−418.9635	−415.8967	−418.9632
Styblinski-Tang	−195.8280	−195.8293	−195.8242

Comparing the results obtained in this section to the results obtained using the stochastic assignment mechanisms, it can be concluded that for the Egg Holder function, Logic C works best among all the methods while Mechanism A works best for the Schwefel 2.26 and the Styblinski-Tang functions.

6.7 Statistical Analysis

A statistical analysis is conducted on the methods presented in this chapter. Two different analysis tools are implemented. The first is evaluating the success rate for each method in solving different problems. The second tool ranks the proposed algorithms using the Sign test [32].

The success rate p_s is computed as $p_s = j_s/n$, where n is the number of runs and j_s is a counter that counts how many times the optimal sequence resulted as an optimal solution at the end of an experiment. **Algorithm** 7.1 shows an outline for this testing algorithm [114]. The $\phi(i)$ is the minimum function value found in the ith run. The threshold tol_f is defined as $tol_f = f(\mathbf{x}_{best}) + \varepsilon$, where $\mathbf{x}_{best}$ is the best known solution to the given problem, $\varepsilon > 0$. The $\psi(i)$ is a vector that includes the minimum-cost solution found in the ith run. The *OptimSeq* is the best known solution.

For each mathematical function presented in **Section** 6.6, the success rate of each logical and stochastic mechanism is calculated numerically. By repeating the same numerical experiment, the obtained solution in each experiment is compared to the best obtained solution and a success rate can be updated as the experiment being repeated. For each of the three logical mechanisms presented in this chapter, the success rates in finding the best solution for the Schwefel 2.26 function is shown in **Fig.** 6.17. Logic B has a success rate of about 70% which is less than that of logics A and C, which is about 100%. For each of the three logical mechanisms presented in this paper, the success rates in finding the best

---

Algorithm 6.2 Success rate algorithm

Apply the HGGA to the problem n times
set $j_s = 0$
for all $i \in 1, ..., n$ **do**
   Compute $\psi(i)$
   Compute $\phi(i)$
   if $\phi(i) < tol_f$ **then**
      $OptimSeq = \psi(i)$
   end if
   if $\psi(i) = OptimSeq$ **then**
      $j_s = j_s + 1$
   end if
end for

---

solution for the Egg Holder function is shown in **Fig.** 6.16. Logic B has a success rate less than that of logics A and C. Both logics A and C settle at a success rate of about 30%. In all three mathematical functions, Logic A and Logic C have very close success rates and Logic B has a lower value for the success rate.

For the stochastic mechanisms presented in this chapter, the success rate of the Schwefel 2.26 function is shown in **Fig.** 6.18. In **Fig.** 6.18, the mutation rate of Mechanism A is 0.001, for Mechanism B is 0.01, and for Mechanism D is 0.03. As shown in **Fig.** 6.18, Mechanisms A, B, and C have higher success rates compared to Mechanism D. Note that Mechanism D has also resulted in higher function values.

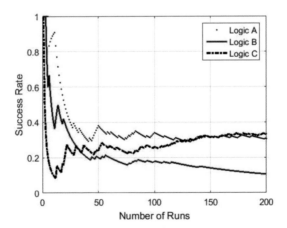

Figure 6.16: Success rate versus number of runs for Egg Holder function using three different logics for tags evolution.

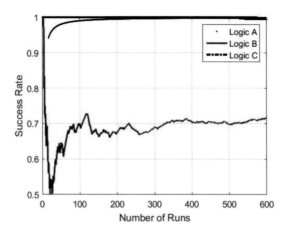

Figure 6.17: Success rate versus number of runs for Schwefel 2.26 function using three different logics for tags evolution.

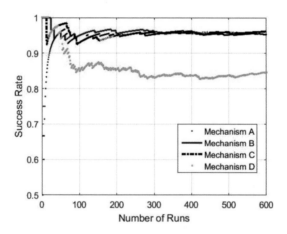

Figure 6.18: Success rate of the Schwefel 2.26 function for different stochastic mechanisms.

The Sign test is a pairwise comparison between different algorithms. It has been applied on the algorithms presented in this chapter. Each algorithm is ranked based on the total number of cases in which the algorithm produces the best function value. **Table** 6.6 summarizes the results obtained in **Section** 6.6, where the best function value obtained using each algorithm is listed. **Table** 6.6 also lists the results obtained using Mechanisms E, F, G, and H which are detailed in reference [4]; they are used here for the purpose of ranking. Listed also in **Table** 6.6 is the solutions obtained using the No-tag HGGA which is originally developed in reference [3].

Table 6.6: Best objective function values of all the test cases obtained from all algorithms.

	Egg Holder	Schwefel 2.26	Styblinski-Tang
Logic A	−3386.2499	−418.9635	−195.8280
Logic B	−3192.6139	−415.8967	−195.8293
Logic C	−3634.1438	−418.9632	−195.8242
Mechanism A	−3692.314081	−418.8967	−195.8306992
Mechanism B	−3599.845154	−418.4088	−195.828109
Mechanism C	−3559.5751	−418.2119	−195.82982
Mechanism D	−3484.255119	−418.1721	−195.8261996
Mechanism E	−3552.8947	−418.2850	−195.8298
Mechanism F	−3644.2279	−418.0799	−195.8278
Mechanism G	−2571.8028	−375.3514	−191.8393
Mechanism H	−2134.7217	−303.5320	−194.5325
Simple HGGA	−2749.2646	−335.1844	−156.6646

Using the data presented in **Table** 6.6, the Sign rank is computed by counting how many times each algorithm resulted in the best function value among all functions. **Table** 6.7 lists the Sign rank for each algorithm. Mechanism A then Logic A have the highest ranks. It is to be noted here, though, that this metric is best when the number of tested functions is high.

Table 6.7: Sign test ranks.

	Sign rank		Sign rank
No-tag HGGA	0	Mechanism A	2
Mechanism B	0	Mechanism C	0
Mechanism D	0	Mechanism E	0
Mechanism F	0	Mechanism G	0
Mechanism H	0	Logic A	1
Logic B	0	Logic C	0

6.8 Markov Chain Model of HGGA

This section presents a convergence analysis that proves HGGAs generate a sequence of solutions with the limit value of the global optima. For an analytical proof, the homogeneous finite Markov models of different mechanisms discussed in this chapter are derived, and the convergence of the HGGAs with tag evolution mechanisms are presented. The optimization problem is considered a maximization problem with strictly positive values for the objective function. In **Section** 6.2.2, a review for the Markov model for binary canonical genetic algorithms (CGAs) was presented and its convergence was analyzed. Here the transition matrix of different HGGA mechanisms are derived. The HGGA is here proved to be convergent, for all the stochastic or logical mechanisms defined in **Section** 6.5. The approach to prove that these HGGA mechanisms are convergent, in general, is as follows:

> First we show that the HGGA can be modeled as a Markov process. Then it is shown that the selection, mutation, and crossover matrices have the properties described in Lemma 1. Therefore, the extended transition matrix of HGGA is reducible and can be written in the form of Eq. (6.9). Finally, Theorem 2 can be used to prove the convergence.

Similar to the canonical GA, any future state of the HGGA population is only dependent on the current population and is independent from the previous history. Hence, if the transition product matrix **CMS** of a HGGA mechanism is stochastic, then the HGGA with that mechanism can be considered as a Markov processes.

To prove that the **CMS** matrix is stochastic and primitive, the intermediate matrices of **C**, **M** and **S** need to be derived. They are derived in this section. It is assumed that the single-point crossover is selected for the genes, unless otherwise stated. The number of genes is L and the number of the tags is L_t. $H(i, j)$ is the Hamming distance between the genes of i and j (number of bits that must be altered by mutation to transform the *genes* of j into the *genes* of i) and is $0 \leq H(i, j) \leq L$. $H_t(i, j)$ is the Hamming distance between the tags of i and j (number of bits that must be altered by mutation to transform the *tags* of j into the *tags* of i) and is $0 \leq H_t(i, j) \leq L_t$. In all the mechanisms, the genes go thorough selection, mutation, and crossover similar to the standard genetic algorithm and only the tags evolution is different.

The transition probability matrices determine the probability of transferring a solution i to solution j; that is to change the L genes of solution i to be the same as the L genes of solution j, and change the L_t tags of solution i to be the same as the L_t tags of solution j.

1. Selection Matrix S The selection operator for the HGGA is not different from that of a canonical GA one. For example, for a fitness proportionate selection, the probability that a solution i is selected only depends on the objective value, which in turn is a function of the values of the genes as well as the values of the tags. Hence, the selection matrix is computed for the HGGA in a similar way to that of the GA as follows.

The probability of selecting a solution $i \in S$, from a population described by the probability distribution vector $\bar{n} \in S'$ is [31]:

$$P_1(i|\bar{n}) = \frac{n(i).R(i)}{\sum_{j \in S} n(j).R(j)} \tag{6.28}$$

where $\bar{n} = (n(0), n(1), ..., n(2^l - 1))$ is the current generation and $n(i)$ represents the number of occurrences of solution i, and $R(i)$ is the objective value for solution i and is strictly positive. Therefore, given the present generation is $\bar{n}$, the conditional probability of the successor generation $\bar{m}$ is a multinomial distribution [31]:

$$P_1(\bar{m}|\bar{n}) = \binom{M}{\bar{m}} \prod_{i \in S} P_1(i|\bar{n})^{m(i)} \tag{6.29}$$

where,

$$\binom{M}{\bar{m}} = \frac{M!}{\prod_{i \in S}(m(i)!)} \tag{6.30}$$

The transition probability matrix of the Markov chain where only the selection operation is applied is $\bar{P} = [P_1(\bar{m}|\bar{n})]$. This matrix is positive, stochastic, and column-allowable. Hence, the transition matrix due to only selection operation in HGGA is stochastic, positive, and column-allowable.

Mutation M and Crossover C Matrices In this section, the explicit formulation of mutation and crossover matrices are derived and it is shown that for all of the mechanisms, the mutation matrix is stochastic and positive and the crossover matrix is stochastic. The general scheme for deriving these matrices is first presented; then followed by its application to each mechanism. Assume a nonzero value for the mutation probability, i.e., $0 < p_m(k) \leq 1/2$. In the mutation operation in the CGA, the probability of transforming j into i can be calculated as $p_m^{H(i,j)}(1 - p_m)^{L-H(i,j)}$. Thus the transition probability, due to both selection and mutation operations, is [31]:

$$P_2(i|\bar{n}) = \sum_{j \in S} p_m^{H(i,j)}(1 - p_m)^{L-H(i,j)} P_1(j|\bar{n})$$

$$= \frac{1}{(1+\alpha)^L} \sum_{j \in S} \alpha^{H(i,j)} P_1(j|\bar{n}), \bar{n} \in S', i \in S \tag{6.31}$$

where $\alpha = \dfrac{p_m}{1 - p_m}$.

$$\therefore P_2(i|\bar{n}) = \frac{\sum_{j \in S} \alpha^{H(i,j)}(n(j).R(j))}{(1+\alpha)^L . \sum_{k \in S} n(k).R(k)} \tag{6.32}$$

The multinomial distribution for $P_2(\bar{m}|\bar{n})$ can be defined as [31]:

$$P_2(\bar{m}|\bar{n}) = \binom{M}{\bar{m}} \prod_{i \in S} P_2(i|\bar{n})^{m(i)} \tag{6.33}$$

Then the transition matrix of selection and mutation would be $\bar{P} = [P_2(\bar{m}|\bar{n})]$. Note that α is positive for $0 < p_m \leq 1/2$. As can be seen from Eq. (6.32), since α is positive, R is positive, and $n \geq 0$, then the $\bar{P}$ matrix is primitive.

Regarding the crossover operation, assume that a single-point crossover is applied. The new function $I(i, j, k, s)$ is defined where $i, j, k \in S$, and $s \in [1, ..., L-1]$ is a bit string. The selected parents are i, j and k is a potential descendant string after a crossover at random location s which is assumed uniformly distributed. If k is produced by crossing i and j at the location s, then $I(i, j, k, s) = 1$, otherwise $I(i, j, k, s) = 0$. The conditional probability of producing k via selection and crossover operations can be derived as [31]:

$$P_2'(k|\bar{n}) = \sum_{i \in S} \sum_{j \in S} \left(P_1(i|\bar{n}).P_1(j|\bar{n}).\frac{p_c}{L-1} \sum_s I(i, j, k, s) \right) + (1 - p_c).P_1(k|\bar{n})$$

(6.34)

Therefore the conditional probability of producing k via selection, mutation, and crossover operations is [31]:

$$P_3(i|\bar{n}) = \frac{1}{(1+\alpha)^L} \sum_{j \in S} \alpha^{H(i,j)} P_2'(j|\bar{n})$$

(6.35)

Then:

$$P_3(\bar{m}|\bar{n}) = \binom{M}{\bar{m}} . \prod_{i \in S} P_3(i|\bar{n})^{m(i)}$$

(6.36)

By inspection of Eq. (6.34) and Eq. (6.35), it can be seen that this three-operator Markov chain is primitive. Then, based on the results of **Section** 6.2.2 this GA model, maintaining the best solution found over time, converges to the global optimum.

Here, the above results are applied to each of the HGGA mechanisms.

◼ **Mechanism A:** In this mechanism, the tags can crossover independently from the genes and there is is a 10% mutation probability in the tags. This implies that the intermediate transition matrix for mutation (**M**) consists of two parts, where the Hamming distance of $H(i, j)$ is the number of bits in the genes only that need to be altered by mutation, and $H_t(i, j)$ is the number of bits in the tags only that need to be altered by mutation. Hence the probability can be described as follows:

$$P_2(i|\bar{n}) = \sum_{j \in S} p_m^{H(i,j)} (1 - p_m)^{L-H(i,j)} p_{mt}^{H_t(i,j)} (1 - p_{mt})^{L_t-H_t(i,j)} P_1(j|\bar{n})$$

(6.37)

Note that the probability that solution j is transfered to solution i is $p_m^{H(i,j)}(1-p_m)^{L-H(i,j)}(0.1)^{H_t(i,j)}(0.9)^{L_t-H_t(i,j)} > 0$ for all $i, j \in S$ when $0 < P_m < 0.5$. Thus, **M** is positive. For the crossover operation:

$$P_2'(k|\bar{n}) = \sum_{i \in S} \sum_{j \in S} \left(P_1(i|\bar{n})P_1(j|\bar{n}) \frac{p_c}{L-1} \frac{1}{L_t-1} \sum_s I'(i,j,k,s,s_t) \right)$$
$$+ (1-p_c)P_1(k|\bar{n})$$

$$(6.38)$$

The $I'(i,j,k,s,s_t)$ takes values $\{0,1\}$, where 1 shows that child k (genes and tags) is produced by the crossover of parents i and j at site s in the genes and at site s_t in the tags. Therefore, the conditional probability of constructing a bit string k via selection, mutation, and crossover operations in HGGA is:

$$P_3(i|\bar{n}) = \frac{1}{(1+\alpha)^{L+L_t}} \sum_{j \in S} \alpha^{H(i,j)} P_2'(j|\bar{n}) \qquad (6.39)$$

Then the transition matrix for Mechanism A can be computed by substituting Eq. (6.39) into Eq. (6.36). Note that L is replaced by $L+L_t$ to account for the additional tags. By inspection of Eq. (6.39), it can be concluded that this transition matrix of HGGA with mechanism A is stochastic and positive.

■ Mechanism B: In this mechanism, the tags are considered as design variables in the crossover operation. The arithmetic crossover is used in this mechanism, where the number of variables in this case is $L+L_t$. Hence, it can be concluded that the crossover transition matrix $P_2'(k|\bar{n})$ (defined in Eq. (6.34)) for mechanism B is stochastic. The mutation operation in mechanism B is similar to that of mechanism A, and hence the mutation transition matrix $P_2(i|\bar{n})$ can be computed using Eq. (6.46) for mechanism B, which is positive when $0 < P_m < 0.5$. Finally, the $P_2(i|\bar{n})$ and $P_2'(k|\bar{n})$ matrices are used to compute $P_3(\bar{m}|\bar{n})$ using Eqs. (6.35) and (6.36). Then the overall transition matrix $P_3(\bar{m}|\bar{n})$ is primitive for mechanism B.

■ Mechanism C: here an arithmetic crossover operation is used for the genes, while the tags are copied from one of the parents as described in **Section** 6.5. The selection and crossover transition probability is defined as follows:

$$P_2'(k|\bar{n}) = \sum_{i \in S} \sum_{j \in S} P_1(i|\bar{n})P_1(j|\bar{n})p_c F_A(i,j,k,\lambda) F_{T_1}(i,j,k,f_{m_1}(i),f_{m_1}(j))$$
$$+ (1-p_c)P_1(k|\bar{n})$$

$$(6.40)$$

where F_A is 1 if the arithmetic crossover of genes in parents i and j, along with the weight coefficient λ result in the genes of solution k; otherwise $F_A = 0$. Also, F_{T_1} is 1 if the tags of solution k are similar to the tags of the parent that has better f_{m_1}; otherwise $F_{T_1} = 0$. For example, if parents i and j are selected and their modified cost values are $f_{m_1}(i)$ and $f_{m_1}(j)$ (defined in **Section** 6.5, Mechanism C), then if $f_{m_1}(i)$ is better than $f_{m_1}(j)$ and the tags of k are similar to the tags of i, then $F_{T_1} = 1$; otherwise $F_{T_1} = 0$. Hence, the resulting crossover probability matrix is stochastic. The Mutation operation is similar to that of mechanisms A and B, and therefore, it is stochastic and positive.

■ Mechanism D: similar to mechanism C, the crossover probability can be written as:

$$P_2'(k|\bar{n}) = \sum_{i \in S} \sum_{j \in S} P_1(i|\bar{n}) P_1(j|\bar{n}) p_c F_A(i,j,k,\lambda) F_{T_2}(i,j,k,f_{m_2}(i),f_{m_2}(j))$$
$$+ (1 - p_c) P_1(k|\bar{n})$$

(6.41)

where F_A is 1 if the arithmetic crossover of genes in parents i and j along with weight the coefficient λ result in the genes of solution k; otherwise $F_A = 0$. Also, F_{T_2} is 1 if the tags of solution k are similar to the tags of the parent that has better f_{m_2}; otherwise $F_{T_2} = 0$. Hence, the resulting crossover probability matrix is stochastic. The Mutation operation is similar to that of mechanisms A and B, and therefore, it is stochastic and positive.

■ Mechanism E: tags evolve through a mutation operation with a certain mutation probability. Let p_{mt} be the mutation probability of the tags, then:

$$P_2(i|\bar{n}) = \sum_{j \in S} p_m^{H(i,j)}(1 - p_m)^{L - H(i,j)} p_{mt}^{H_t(i,j)}(1 - p_{mt})^{L_t - H_t(i,j)} P_1(j|\bar{n})$$

(6.42)

which is stochastic. Also since p_m and p_{mt} are positive and less than 0.5, then $P_2(i|\bar{n})$ is positive. The crossover is only applied to the genes in this mechanism, hence:

$$P_2'(k|\bar{n}) = \sum_{i \in S} \sum_{j \in S} \left(P_1(i|\bar{n}) . P_1(j|\bar{n}) . \frac{p_c}{L-1} \sum_s I(i,j,k,s) \right)$$
$$+ (1 - p_c) . P_1(k|\bar{n})$$

(6.43)

Similar to the CGA, the matrix $P_2'(k|\bar{n})$ above is stochastic.

■ Mechanism F: In this mechanism, the tags are considered as discrete variables similar to the design variables in the chromosome. The crossover and mutation operations are performed on all the variables (genes and tags). The mutation transition probability is then as follows:

$$P_2(i|\bar{n}) = \sum_{j \in S} p_m^{H(i,j)+H_t(i,j)} (1 - p_m)^{L+L_t-H(i,j)-H_t(i,j)} P_1(j|\bar{n}) \qquad (6.44)$$

which results in a positive and stochastic mutation matrix. Also the stochastic crossover transition probability can be calculated as follows:

$$P_2'(k|\bar{n}) = \sum_{i \in S} \sum_{j \in S} \left(P_1(i|\bar{n}).P_1(j|\bar{n}).\frac{p_c}{L+L_t-1} \sum_{s} I(i,j,k,s) \right) \\ + (1 - p_c).P_1(k|\bar{n}) \qquad (6.45)$$

- Mechanism G: In this mechanism, the tags are considered as discrete variables similar to the design variables in the chromosome; yet only the crossover operation is applied to the tags. Since there is no mutation in the tags, the mutation transition probability is as follows:

$$P_2(i|\bar{n}) = \sum_{j \in S} p_m^{H(i,j)} (1 - p_m)^{L+L_t-H(i,j)} P_1(j|\bar{n}) \qquad (6.46)$$

which results in a positive and stochastic mutation matrix. The stochastic crossover probability matrix is similar to Eq. (6.45).

- Mechanism H: In this mechanism, the tags are considered as discrete variables similar to the design variables in the chromosome; yet only the mutation operation is applied to the tags. Hence, the mutation matrix is similar to Eq. (6.44) which is stochastic and positive. The crossover probability matrix is similar to Eq. (6.43); which is stochastic.

Let the length of the alleles be $2L_t$, and let H_a be the Hamming distance between the tags of the i and j alleles (number of bits that must be altered by mutation to transform the tags of j into the tags of i). The maximum of H_a is $2L_t$. Since all the bits go through mutation with probability p_m, the mutation conditional probability can be calculated as:

$$P_2(i|\bar{n}) = \sum_{j \in S} p_m^{H(i,j)+H_a(i,j)} (1 - p_m)^{L+2L_t-H(i,j)-H_a(i,j)} P_1(j|\bar{n}) \qquad (6.47)$$

which results in a stochastic and positive mutation matrix. There are two crossover points, one in the genes and one in the tags such that $s_t \in [1, ..., L_t - 1]$. The crossover points in tags (s_t) are similar in the dominant and recessive alleles. Hence:

$$P_2'(k|\bar{n}) = \sum_{i \in S} \sum_{j \in S} \left(P_1(i|\bar{n}).P_1(j|\bar{n}).\frac{p_c}{L-1} \cdot \frac{1}{L_t-1} \sum_{s} I'(i,j,k,s,s_t) \right) \\ + (1 - p_c).P_1(k|\bar{n}) \qquad (6.48)$$

where $I'(i, j, k, s, s_t)$ is 1 if the crossover of i and j at site s in genes and site s_t in tags produce k, otherwise $I'(i, j, k, s, s_t) = 0$. The crossover matrix in Eq. (6.48) is stochastic.

■ Logic A: the member of the current generation ($\bar{n}$) is split into two groups of equal size. For the first group, the Hidden-Or logic is applied on the tags and for the other half, the Active-Or logic is used in the tags. There is no mutation in the tags; hence the mutation probability matrix is defined as in Eq. (6.46). Let F_{HO} and F_{AO} be functions that can have values of 0 or 1. If the Hidden-Or operator on the tags of i and j results in the tags of k, then $F_{HO}(i, j, k) = 1$, otherwise $F_{HO}(i, j, k) = 0$. If the Active-Or operator on the tags of i and j results in the tags of k, then $F_{AO}(i, j, k) = 1$, otherwise $F_{AO}(i, j, k) = 0$. For the first half of the children the crossover probability matrix is then:

$$P_2'(k|\bar{n}_1) = \sum_{i \in S} \sum_{j \in S} P_1(i|\bar{n}_1).P_1(j|\bar{n}_1).F_{HO}(i, j, k).\frac{p_c}{L-1} \sum_s I(i, j, k, s)$$
$$+ (1 - p_c).P_1(k|\bar{n}_1)$$

(6.49)

and for the second half of the children:

$$P_2''(k|\bar{n}_2) = \sum_{i \in S} \sum_{j \in S} P_1(i|\bar{n}_2).P_1(j|\bar{n}_2).F_{AO}(i, j, k).\frac{p_c}{L-1} \sum_s I(i, j, k, s)$$
$$+ (1 - p_c).P_1(k|\bar{n}_2)$$

(6.50)

Where $\bar{n}_1$ represents one half of the GA search space, and $\bar{n}_2$ represents the other half of the GA search space. The conditional probability of producing k with i and j via selection and crossover is $P_2'(k|\bar{n}_1) \times P_2''(k|\bar{n}_2)$, which results in a stochastic matrix.

■ Logic B: The Hidden-OR logic is used for both children. Even though the tags will be the same in both children, the two children represent two different solutions because they have different gene values. There is no mutation for the tags, hence, the mutation probability matrix is defined as in Eq. (6.46). The crossover probability matrix is:

$$P_2'(k|\bar{n}) = \sum_{i \in S} \sum_{j \in S} P_1(i|\bar{n}).P_1(j|\bar{n}).F_{HO}(i, j, k).\frac{p_c}{L-1} \sum_s I(i, j, k, s)$$
$$+ (1 - p_c).P_1(k|\bar{n})$$

(6.51)

Both mutation and crossover matrices are stochastic; in addition the mutation conditional probability is positive.

■ Logic C: The Active-OR logic is used for both children. Even though the tags will be the same in both children, the two children represent two different solutions because they have different gene values. The mutation probability matrix is defined as in Eq. (6.46). The crossover probability matrix is:

$$P_2'(k|\bar{n}) = \sum_{i \in S} \sum_{j \in S} P_1(i|\bar{n}).P_1(j|\bar{n}).F_{AO}(i,j,k).\frac{p_c}{L-1} \sum_s I(i,j,k,s)$$
$$+ (1 - p_c).P_1(k|\bar{n})$$

(6.52)

Both mutation and crossover matrices are stochastic; in addition the mutation conditional probability is positive.

By calculating the **C**, **M**, and **S** matrices of different mechanisms, we can now continue on the convergence analysis. As shown, the mutation matrices in all the mechanisms are stochastic and positive. The selection matrix is also stochastic and positive; and hence it is column-allowable. Also the crossover matrices are stochastic. Hence, the **CMS** matrix is positive (Lemma 1). Since the HGGA maintains the best solution found over time after selection, Theorem 6 can be used to prove that all mechanisms of HGGA presented above are convergent.

6.9 Final Remarks

The HGGA presented in this chapter demonstrated success in searching for optimal solutions in VSDS problems, for a range of applications as detailed in the Applications section. Few other algorithms in the literature can perform search for the optimal architecture in a VSDS system. For example references [19] and [63] present algorithms that search for the optimal structural topology in truss and frame structures, respectively. While the algorithms in [19] and [63] are problem specific, they do demonstrate the capability of searching for the optimal solution in a VSDS problem. The dissertation in [121] presents a study on topology optimization of nanophotonic devices and makes a comparison between the homogenization method [109] and genetic algorithms [62].

The closest to HGGA is probably the concept of using null values in genetic algorithms to represent scenarios where the optimizer skips over the null values and construct a solution only from the values that do not have a null value [37, 83, 39]. This concept is similar to the junk genes in the field of genetics, where DNA chromosomes contain some genes that do not code for any traits and are skipped over when the chromosome is read [37]. This algorithm was developed for space trajectory planning using evolutionary algorithms. In particular it enables the optimization of the planetary fly-by sequence. In that regard, the numbers 015 are chosen to code for each body and the null body because they can easily be represented in a binary string with four bits for each potential flyby. For example, consider a mission that departs from Earth and ends at Jupiter. The outer-loop optimizer might generate the following binary string: [0010001110110011]. This string is then converted to integers, yielding [2 3 11 3]. Using a saved coding for the planets, the integers are converted into the sequence VEnE where "n" is a null flyby. Hence the mission sequence in this example is EVEEJ. As can be seen from the previous example, the main difference between the null value approach and the HGGA is that the latter assumes values for the hidden genes, despite not using them when the genes are hidden. These unused values of the hidden genes go through the normal evolution operations of genetic algorithms and might become active in subsequent generations. This feature is not there when representing a no fly-by event by a null value.

Chapter 7

Structured Chromosome Genetic Algorithms

This chapter presents the Structured-Chromosome Evolutionary Algorithm (SCEA) framework that is developed to handle variable-size design space optimization problems. In this framework, a solution (chromosome) is represented by a hierarchical data structure where the genes in the chromosome are classified as dependent and non-dependent genes. This structure provides the capability to apply genetic operations between solutions of different lengths. The SCEA features new concepts for the chromosome structure where the genes are tagged and arranged as a hierarchical linked list. The hierarchy was designed to represent "dependent" genes whose existence depends on the value of other genes in the chromosome. This structure allows flexibility in the number of design variables without reserving space in advance for unused or hidden genes. This arrangement also enables the results of the genetic operations to always yield valid solutions without using hidden genes. Both the Structured-Chromosome Genetic Algorithms (SCGA) and the Structured-Chromosome Differential Evolution Algorithm (SCDE) are presented in this chapter. Using the SCEA in interplanetary trajectory optimization enables the automatic determination of the number of swing-bys, the planets to swing-by, launch and arrival dates, and the number of DSMs as well as their locations, magnitudes, and directions, in an optimal sense. This new method is applied to several interplanetary trajectory design problems. Results show that solutions obtained using this tool match known solutions for some complex problems. A comparison between genetic algorithms and differential evolution in the structured-chromosome framework is presented. The SCGA was also implemented in satellite tracking in [56] where it is used to define observation campaigns for tracking space objects from a network of tracking stations. In the field of commercial aircraft design, high-Lift devices are components located on the aircrafts wing that aim to increase the lift force produced by the wing during slow flight phases, mainly take-off and landing. In the design optimization of these high-Lift devices, the configuration type (number of airfoil elements), along with their positions and shapes are design variables. This optimization problem was solved using the SCGA in [57]. The SCGA was also used in spacecraft design optimization in [55]. Spacecraft design optimization is a complex problem characterized by its mixed-variable problem formulation involving continuous, integer and categorical variables. It is a VSDS problem because some of the variables are included in the problem formulation only if certain technologies are used in the spacecraft. The work in [55] compares the use of SCGA in spacecraft design to the Multi-Population Adaptive Inflationary Differential Evolution Algorithm. This chapter was originally published in [87]. **Section** 7.1 introduces the SCEA chromosome structure along with an explanation of the evolutionary algorithms implemented. **Section** 7.2 details the application of the SCEA to the trajectory optimization problem and presents numerical results.

7.1 Structured-Chromosome Evolutionary Algorithms (SCEAs)

In standard evolutionary algorithms, a chromosome (solution) consists of a single string of genes, all at one level. In the SCEA framework, a chromosome is represented by a hierarchical linked list data structure. Each feature is represented by a gene, and all the genes are linked together to form a complete chromosome. To highlight the difference between the proposed method and the standard EAs, consider **Figs.** 7.1(a) and 7.1(b). A standard EA chromosome is shown in **Fig.** 7.1(a) where genes are stacked together in one string. Different genes are not linked, and hence the number of genes is independent from the value of any gene in the chromosome. The SCEA chromosome structure is shown in **Fig.** 7.1(b) where the genes are organized in different layers. The value of the gene B determines how many genes exist in the second layer (2 genes D and E). Nodes D, E, and F are called dependent nodes. **Figure** 7.2 shows a portion of the structured chromosome for the trajectory optimization problem, with three layers. The first layer contains four variables where the first and last variables (source planet and number of gravitational assist maneuvers) have dependent variables.

Every gene (feature) contains four important fields; these fields are *value*, *type*, *next* and *child*. *Value* is used to hold the gene's actual value; it can be a real or an integer number. *Type* is used to identify the feature type. *Next* is a pointer that points to the immediate neighbor gene at the same level. *Child* is a pointer that points to a gene at the next level as shown in **Fig.** 7.2. Genes that are pointed by child pointers are called dependent genes. In **Fig.** 7.2, square boxes represent

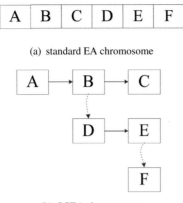

(a) standard EA chromosome

(b) SCEA chromosome

Figure 7.1: Multi-Layer chromosome in SCEA versus one layer chromosome in standard EA.

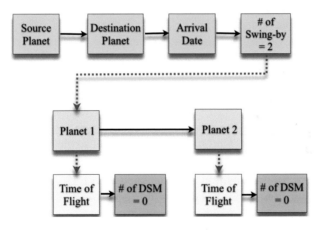

Figure 7.2: Section of a multi-layer chromosome in SCEA.

genes, solid arrows represent *Next* pointers, and dotted arrows represent *Child* pointers.

Two types of evolutionary algorithms are adopted to implement the structured chromosomes approach: the Structured-Chromosome Genetic Algorithm (SCGA) and the Structured-Chromosome Differential Evolution (SCDE). Differential evolution has been successfully used in solving the discrete and mixed-integer space trajectory optimization problems [89, 2]. Each of the two algorithms adopts the standard high level steps from the prospective evolutionary algorithm. Specifically, both algorithms start by initializing the population and run a loop consisting of parent selection, offspring generation, and creation of a new population. The high-level algorithms are shown in **Algorithms** 7.1 and 7.2, where the fitness (F) of a gene is defined as the inverse of its cost (Δv_T):

$$F = \frac{1}{\Delta v_T} \tag{7.1}$$

For offspring generation, the Structured-Chromosome Genetic Algorithm (SCGA) framework uses crossover and mutation operations, while the Structured-Chromosome Differential Evolution (SCDE) framework uses a transformation operation as explained in the following sections.

Algorithm 7.1 Structured Chromosome Genetic Algorithm

Create a population of n chromosomes.
while a termination condition is not satisfied **do**
    for all $i \in 1, ..., n$ **do**
        Select two parent chromosomes
        Produce two offspring using crossover
        Mutate the offspring
        Place the offspring in the new population
    end for
    Replace the population with the new population
end while

Algorithm 7.2 Structured Chromosome Differential Evolution Algorithm

---

Create a population of n chromosomes.
while a termination condition is not satisfied **do**
    for all target chromosomes **do**
        Select three additional parent chromosomes
        Produce one offspring using transformation
        Replace the target chromosome with the offspring if the fitness threshold
        is met
    end for
end while

7.1.1 Crossover in SCGA

The standard crossover operation cannot be used in SCGA. Crossover is an operation that exchanges genes between two different chromosomes (parents) to produce two offspring (children). **Figure** 7.3 shows the traditional concept of crossover where parts of the chromosomes are swapped among parent chromosomes to produce the offspring. Since all chromosomes are of the same length

Figure 7.3: Crossover operation in standard GA.

and are one-layer, the two swapped parts represent the same variables, with different values for these variables. In SCGA, however, the chromosomes are multi-layer and are different from one solution to another. Hence swapping parts among parent chromosomes in SCGA may result in swapping genes that represent different variables; this would result in offspring that are meaningless or non-feasible. For instance, some variables such as the launch/arrival dates, time of light of each leg, time of each DSM, amount of impulses of DSMs, and gravitational assist maneuvers' heights and plane angles are real-values variables, whereas variables such as the number of gravitational assist maneuvers, gravitational assist planets' IDs, and the number of DSMs are integers. A meaningful crossover operation should at least preserve this property for each variable. This chapter introduces a definition for crossover operation that works in the SCGA formulation and produces meaningful offspring.

In SCGA, the proposed crossover operation is only permitted on genes with the same gene *type*. This constraint is placed to guarantee a semantically correct crossover where the *value* fields as well as the dependent genes are swapped. **Figure** 7.4 illustrates the crossover operation. At the initial stage, the first chromosome contains genes A, B, C, D, E and F and second chromosome contains genes A', B', C' and F'. The genes that have the same name have the same *type*. The first exchange is between genes A and A'. This exchange requires swapping only the *value* fields because neither gene has dependent variables. The second exchange is between genes B and B'. This exchange involves two genes with dependent variables, therefore both the value and child fields are swapped. Notice that at the initial stage the length (total number of genes) of the first chromosome is six and the length of second chromosome is four but after the second exchange, the lengths change to four and six, respectively. The last exchange is between genes F and F'. Even though both genes have been moved at the second exchange, they are still valid for another exchange because they were not directly involved in the exchanging process previously.

7.1.2 Mutation in SCGA

The mutation operation allows a gene within a chromosome to take a new value. Each gene has a predefined mutation probability. When a gene is selected for mutation, the software makes sure that the randomly generated new value is consistent with the gene's *type*, e.g., real, integer. The generated value must fall

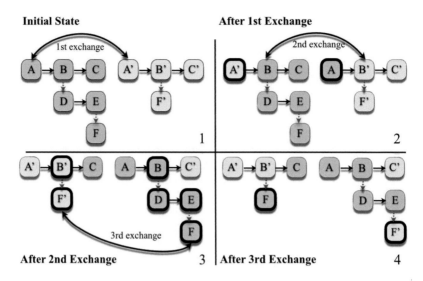

Figure 7.4: Example of crossover operation.

between the lower and upper bounds provided for the gene. If the transformed value is a gene that has dependent children, then the dependent genes are created in a way that ensures semantic correctness. For example, if the number of gravitational assist value decreases, then the extra children are deleted from the tree. Similarly, if the number of gravitational assist value increases, then the existing children are copied to the new chromosome and randomly initialized children are added as additional dependents.

Each gene is represented as a node in the chromosome tree. During the crossover and mutation operations, each gene/node in the candidate chromosome tree is visited once. The process of visiting each node in a tree is called tree traversal. One of the simplest ways to implement tree traversal is to use recursive algorithms. However, recursive algorithms incur significant overhead such as the function call stack and memory allocation [34]. Therefore, we switched to an iterative algorithm that uses stacks. However, we found that stack operations such as push and pop introduce significant overhead given that they have to be invoked millions of times. To gain efficiency, we switched to an iterative algorithm that does not involve stacks [34].

7.1.3   *Transformation in SCDE*

The differential evolution implementation adopted in this chapter uses a single transformation operator that combines mutation and crossover. Transformation starts by selecting a target chromosome, tgt, and three other chromosomes $a, b, c,$

and uses two steps to generate an offpring. The first step is to create a temporary chromosome, *tmp*, using Eq. (7.2).

$$tmp = a + w \times (b - c) \tag{7.2}$$

The w value is the differential weight that adjusts the influence of $(b - c)$ in the new chromosome. In our experiments we used 0.8 for w. The final step is to generate the offpring by crossover from the target chromosome, *tgt*, and the newly generated chromosome, *tmp*.

During the experiment runs we noticed that the $(b - c)$ value frequently yields zero during mutation if the population contains many chromosomes with the same value for gravitational assist maneuvers. This phenomenon causes the population to get stuck at a single gravitational assist value and prevents other possibilities from being explored. To remedy this situation, we use the genetic algorithm mutation operation if $(b - c) = 0$. Otherwise, we use the above differential evolution formula.

7.1.4 Niching in SCGA and SCDE

Previous experience [49] shows that better solutions can be found by implementing a niching function in the optimization algorithm. When applied to a solution, the niching function applies a fitness degradation to that solution's fitness, such that fitness values are depressed in regions where solutions have already been found. This prevents the optimization algorithm from getting trapped in a small region. This chapter implements a sequential niching technique [25]. This technique maintains a "black list" of solutions as shown in **Algorithm** 7.3. If a solution is found to be the best solution in several iterations in a row, it is stored in a "black list". The algorithm is then forced to search away from them so that it can explore other areas in the design space; this is done by assigning a degraded fitness value to each of the "black list" solutions. The "black list" is cleared if the number of black listed solutions reaches a maximum.

In **Algorithm** 7.3, the *best solution age* indicates the number of times a solution with that fitness was found. A *best solution* that has an age greater than the *solution age threshold* is treated as a new local optimum that is blacklisted. The *cost difference threshold* is used to prevent resetting the *solution age* by solutions that have insignificant improvement over the current best solution.

The niching algorithm derates every solution that is sufficiently close to one of the best solutions in the black list. The modified fitness function $F'(sol)$ for a solution is computed by multiplying the original fitness function with a series of derating functions, G. Every time a new blacklisted solution, $blist_{iter}$, is found, the modified fitness function is updated to:

$$F'_{iter+1}(sol) = F'_{iter} \times G(sol, blist_{iter}) \tag{7.3}$$

where $F'_0(sol) = F(sol)$, and $F(sol)$ is the original fitness function.

Algorithm 7.3 The niching algorithm

Set *best solution* to `null`
if *best solution* is `null` **then**
  set the current solution as *best solution*
  set *best solution age* as 1
end if
if distance between current solution cost and best solution cost ¿ *cost difference threshold* **then**
  set the current solution as the *best solution*
  set *best solution age* as 1
end if
if *best solution age* ¿ *solution age threshold* **then**
  put *best solution* in *black list*
  Set *best solution* to `null`.
end if

The derating function, G, used here is defined as follows:

$$G(sol, blist_{iter}) = \begin{cases} 0 & \text{if } dist(sol, blist_{iter}) \leq niche\ radius \\ 1 & \text{otherwise} \end{cases} \quad (7.4)$$

where *niche radius* is a real number used to define local minimum area centered at $blist_{iter}$. The distance function, *dist*, computes the Euclidean distance between two solutions:

$$dist(sol, blist_{iter}) = \sqrt{\sum_{u=1}^{K'} (sol^u - blist_{iter}^u)^2} \quad (7.5)$$

where sol^u and $blist^u$ are u^{th} features of *sol* and $blist_{iter}$ respectively. Note that the number of parameters in solutions *sol* and $blist_{iter}$ might be different because the chromosomes in the population have verying lengths. Therefore, we use the maximum number of features ($K' = \max\{\|sol\|, \|blist_{iter}\|\}$) and if any feature does not exist in sol^u or $blist^u$, we set the missing feature value to zero.

7.2 Trajectory Optimization using SCEA

The number of genes in the Multi Gravity-Assist trajectory with Deep Space Maneuvers (MGADSM) problem is closely tied to its genes' values. The structured chromosome was designed to accomodate genes that may or may not exist depending on the value of other genes. The SCEA guarantees that all of the offspring generated by evolutionary operations are valid solutions to the problem. **Figure** 7.5 shows the chromosome structure of the MGADSM problem. There

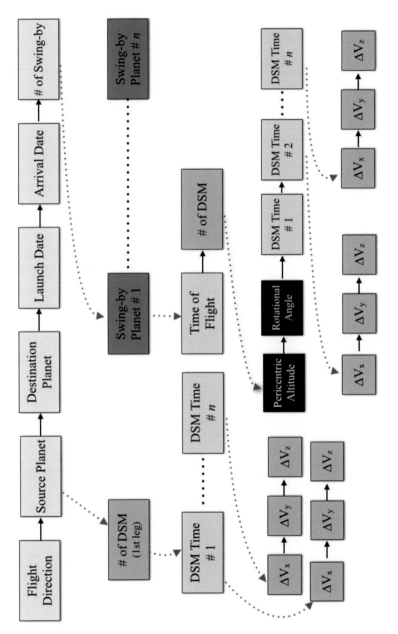

Figure 7.5: Structure of a MGADSM chromosome in SCEA.

are two types of genes: non-dependent and dependent. The non-dependent genes are flight direction, source planet, destination planet, launch date, arrival date, and number of gravitational assist manuevers. The remaining genes are dependent, meaning that they may or may not exist depending on the values of the non-dependent genes. Notice that some subtrees are repeated depending on their parent genes. For example, if the first leg has three DSM, the value of "# of DSM" under "source planet" will be three and there will be three subtrees with DSM time, ΔV_x, ΔV_y, and ΔV_z.

In the all the numerical experiments presented in this chapter, the interplanetary trajectory optimization problem was divided into two stages; the first stage is a zero-DSM stage where the solution assumes no DSMs in the trajectory to find the optimal sequence of gravity assist maneuvers. In the second stage, the sequence is fixed and DSMs are added to the trajectory to refine it. The zero-DSM assumption is not the best way to solve this problem since it might unfairly penalize good solutions. However, the zero-DSM assumption was a necessary shortcut for this particular version of the method to avoid the excessive high computational cost.

In order to test the SCEA algorithm, statistics were generated on its behavior over a number of runs and for a given computational effort measured in terms of the number of function evaluations. The algorithm presented by Vasile et al. [114] was implemented to measure the success rate of an algorithm in finding the optimal mission cost. The test algorithm measures how many times the best solution is obtained over a number of runs. The best solution, however, gets updated every time a better solution is obtained. So, the success rate after the first run is 1 and gets tuned as more runs are performed. The steady state value of the success rate is then the measure for the algorithm efficiency. Because the main advantage of the SCEA is its capability to automatically find the optimal sequence of planets, another measure is also implemented to measure the success rate of the algorithm in finding the optimal sequence of gravitational assist maneuvers. The success rate SR is computed as $SR = SC/NR$, where NR is the number of runs and SC is a counter that counts how many times the optimal sequence resulted as an optimal solution at the end of an experiment.

The implementation was done using Matlab R2011b. The experimental runs were performed on a Linux CentOS 5.5 cluster with 32 nodes having Intel Xeon CPU5120s at 1.86GHz. A search was terminated when a predefined number of cost function evaluations was met. For statistical analysis, each experiment was run 200 times.

7.2.1 Earth-Mars Mission

The problem of finding the minimum cost MGADSM trajectory for the Earth-Mars (EM) mission was solved using SCEA. The total number of design variables for this mission is 33 (7 Discrete design Variable (DDVs) and 26 Contin-

Table 7.1: Bounds of design variables for Earth-Mars mission.

Design Variables	Lower Bound	Upper Bound
No. of gravitational assist manuevers	0	2
gravitational assist planets	1 (Mercury)	8 (Neptune)
No. of DSMs per leg	0	1
Flight direction, f	Posigrade	Retrograde
Departure date, t_d	01-Jun-2004	31-Jul-2004
Arrival date, t_a	01-Apr-2005	01-Jul-2005
Time of flight (days)/leg, $T_1,..., T_i$	30	300
gravitational assist normalized pericenter altitude, $\bar{h}_1,..., \bar{h}_i$	0.1	10
gravitational assist plane rotation angle (rad), $\eta_1,..., \eta_i$	$-\pi/6$	2π
Epoch of DSM, $\varepsilon_1,..., \varepsilon_j$	0.05	0.95
DSM (km/s), $\Delta v_1,..., \Delta v_k$	-5	5

Table 7.2: MGADSM solution trajectory for the EVM mission.

Mission Parameter	MGADSM Model
Departure date	05-Jun-2004 17:00:29
Departure impulse, km/s	4.6127
Venus gravitational assist date	20-Nov-2004 11:26:28
Arrival date	13-May-2005 19:13:00
Arrival impulse, km/s	6.1661
Time of flight between gravitational assist maneuvers (days)	167.8/174.2
gravitational assist normalized pericenter altitude, $\bar{h}$	3.4
Mission Cost (km/s)	**10.7788**

uous design Variable (CDVs)). The lower and upper bounds for all the design variables are listed in **Table** 7.1. The maximum possible number of gravitational assist manuevers was selected to be two. The maximum number of DSMs in each leg was selected to be two impulses. The direction of flight can be posigrade or retrograde. The time of flight for each leg except the last one was selected between 40 and 300 days. A population of 100 individuals was used, and the maximum number of cost function evaluations was set to 200,000. The resulting trajectory has a single gravitational assist manuever at Venus (EVM where E = Earth, V = Venus, and M = Mars), with a total cost of 10.7788 km/s, including the initial hyperbolic departure provided by the launch vehicle. The fittest trajectory has no DSM as shown in **Table** 7.2. Statistical analysis was conducted by running the same numerical experiment 200 times. **Figure** 7.6 shows the success rates for SCDE and SCGA. The success rate for both algorithms is 100% in finding the optimal gravitational assist planet sequence (EVM). The success rates for finding the optimal cost settles at above 95% for both SCDE and SCGA.

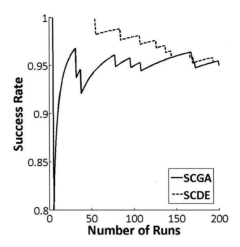

Figure 7.6: SCGA and SCDE success rates in finding the best cost for the Earth-Mars mission.

7.2.2 *Earth-Saturn Mission (Cassini 2-like Mission)*

The Cassini 2 trajectory is one of the most complicated multi gravity assist trajectories[1]. The mission's target is to study planet Saturn and its moons. This section presents the results obtained for a Cassini-2 like mission; the difference between the solved case and the actual Cassini-2 problem is in the bounds on the variables. There are 33 independent design variables in this problem (11 DDVs and 22 CDVs). **Table** 9.7 presents the upper and lower bounds for the design variables. Both the SCGA and the SCDE are used to search for a minimum-cost trajectory for an Earth-Saturn trip. A population of 20 individuals was used and the maximum number of cost function evaluations was selected to be 300,000.

The optimization process was divided into two steps to reduce the computational cost. First, a zero-DSM solution was sought to determine the planet sequence. The resulting zero-DSM planet sequence was a three gravitational assist trajectory EVEJS (J = Jupiter and S = Saturn) with a cost of 10.12 km/s. In the second step, the problem was solved using the full MGADSM formulation assuming that the mission scenario is EVEJS as obtained from the first step. The result of the second optimization step was a single DSM of 694.5 m/s in the second leg, as shown in **Table** 7.4. The total transfer cost was reduced to 9.95 km/s in the second optimization step. **Table** 7.4 shows the zero-DSM trajectory as well as the final trajectory. As shown in **Table** 7.4, the effect of the DSM is reducing the departure impulse. The resulting trajectory is shown in **Figs.** 7.7(a) and 7.7(b).

---

[1]European Space Agency, GTOP act trajectory database, July 2009, http://www.esa.int/gsp/ACT/inf/op/globopt.htm

Table 7.3: Bounds of Earth-Saturn mission's design variables.

Design Variables	Lower Bound	Upper Bound
No. of gravitational assist maneuvers	1	4
gravitational assist planets	2 (Venus)	5 (Jupiter)
No. of DSMs per leg	0	1
Flight direction, f	Posigrade	Retrograde
Departure date, t_d	01-Nov-1997	31-Nov-1997
Arrival date, t_a	01-Jan-2007	30-Jun-2007
Time of flight (days)/leg, $T_1,..., T_i$	40	1000
gravitational assist normalized pericenter altitude, $\bar{h}_1,..., \bar{h}_i$	0.1	10
gravitational assist plane rotation angle (rad), $\eta_1,..., \eta_i$	$-\pi/6$	2π
Epoch of DSM, $\varepsilon_1,..., \varepsilon_j$	0.05	0.95
DSM (km/s), $\Delta v_1,..., \Delta v_k$	-5	5

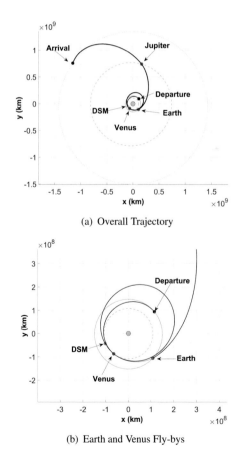

(a) Overall Trajectory

(b) Earth and Venus Fly-bys

Figure 7.7: Solution trajectory for the Cassini 2-like mission.

Table 7.4: Obtained trajectory of Earth-Saturn mission (EVEJS) using both SCGA and SCDE.

Mission Parameter	Zero-DSM Model	MGADSM Model
Departure date	01-Nov-1997 00:00:00	01-Nov-1997 00:00:00
Departure impulse, km/s	4.0505	3.7877
Venus gravitational assist date	28-Mar-1998 18:57:29	29-Mar-1998 11:47:31
DSM date	–	27-May-1999 03:04:39
DSM impulse, km/s	–	0.6945
Earth gravitational assist date	05-Aug-1999 14:59:46	08-Aug-1999 02:43:09
gravitational assist maneuver cost	1.8686	1.2556
Jupiter gravitational assist maneuver date	02-Apr-2001 06:22:18	02-Apr-2001 23:53:32
Arrival date	29-Jun-2007 23:59:59	30-Jun-2007 00:00:00
Arrival impulse, km/s	4.2041	4.2073
Time of flight between gravitational assist maneuvers (days)	147.8/494.8/605.6/2279.7	148.5/496.6/603.9/2279.0
gravitational assist normalized pericenter altitudes	1.6/2.1/1.4	1.6/2.1/1.3
Total Time of flight (days)	3527	3528
Mission Cost (km/s)	**10.1232**	**9.9450**

Another good solution was obtained that has slightly lower cost than the above solution but one extra gravitational assist maneuver. This solution has the four gravitational assist sequence EVVEJS and a total cost of 9.44 km/s. The details of this solution are listed in **Table** 7.5.

Table 7.5: Obtained trajectory of Earth-Saturn mission (EVVEJS).

Mission Parameter	MGADSM Model
Departure date	01-Nov-1997 11:53:29
Departure impulse, km/s	4.0696
Venus gravitational assist maneuver date	08-May-1998 22:46:43
DSM date	12-Aug-1998 11:17:48
DSM impulse, km/s	0.64346
Venus gravitational assist maneuverdate	30-Jun-1999 08:50:49
gravitational assist maneuver cost	0.4703
Earth gravitational assist maneuver date	20-Aug-1999 10:09:57
gravitational assist maneuver cost	0.0064
Jupiter gravitational assist maneuver date	31-Mar-2001 03:45:29
gravitational assist maneuver cost	0.0018
Arrival date	03-Apr-2007 00:51:43
Arrival impulse, km/s	4.2497
Time of flight between gravitational assist maneuvers (days)	188.4/417.4/51/588.7/2193.9
Total Time of flight (days)	3440
Mission Cost (km/s)	**9.4412**

Statistical analysis was conducted on this problem using SCGA and SCDE. after running each method 200 times. The obtained results are shown in **Figs.** 7.8 and 7.9. **Figure** 7.8 shows the success rates for the first optimization step in finding the planet sequence. **Figure** 7.9 shows the success rates for the second optimization step in finding the best cost. These results confirm that the SCDE has higher success rate than the SCGA for the Earth-Saturn mission despite the fact that both methods found the same solution.

7.2.3   *Jupiter Europa Orbiter Mission*

The SCEAs were used to investigate a minimum cost MGADSM trajectory for the Jupiter Europa Orbiter Mission (JEO) that was planned to take place in the decade 2018-2028 [40] (the mission has been canceled). The total number of design variables in this problem is 33 (7 DDVs and 26 CDVs). Wide ranges were allowed for the design variables. **Table** 7.6 lists the lower and upper bounds of the associated design variables. A population of 30 individuals was used and the maximum number of cost function evaluations was set to 300,000. The problem

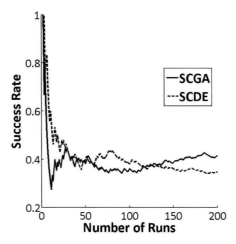

Figure 7.8: SCGA and SCDE success rates in finding the planet sequence for the Cassini 2-like mission.

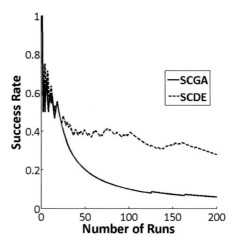

Figure 7.9: SCGA and SCDE success rates in finding the best cost for the Cassini 2-like mission.

is solved in two steps, the Zero-DSM step then the MGADSM step. The resulting fittest scenario is a three gravitational assist trajectory around Venus, Earth, and Earth (EVEEJ), as shown in **Figs.** 7.10(a) and 7.10(b). The details of the resulting solution is listed in **Table** 7.7. The solution is a posigrade multi-revolution trajectory. The cost of the resulting solution is 8.93 km/sec. The resulting solution has one DSM.

Statistical analysis was conducted on the JEO mission for both SCDE and SCGA. **Figures** 7.11(a) and 7.11(b) show the statistical analysis for the SCGA

Table 7.6: Bounds of Earth-Jupiter mission's design variables.

Design Variables	Lower Bound	Upper Bound
No. of gravitational assist maneuver	1	4
gravitational assist planets	2 (Venus)	4 (Mars)
No. of DSMs per leg	0	1
Flight direction, f	Posigrade	Retrograde
Departure date, t_d	31-Jan-2018	31-Dec-2020
Arrival date, t_a	January 1 2026	December 31 2028
Time of flight (days)/leg, $T_1,..., T_i$	100	1000
gravitational assist normalized pericenter altitude, $\bar{h}_1,..., \bar{h}_i$	0.1	10
gravitational assist plane rotation angle (rad), $\eta_1,..., \eta_i$	$-\pi/6$	2π
Epoch of DSM, $\varepsilon_1,..., \varepsilon_j$	0.05	0.95
DSM (km/s), $\Delta v_1,..., \Delta v_k$	-5	5

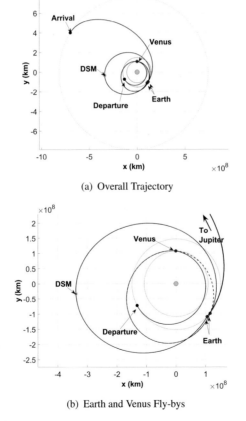

(a) Overall Trajectory

(b) Earth and Venus Fly-bys

Figure 7.10: The best found solution trajectory for the JEO mission.

Table 7.7: Trajectory of Earth-Jupiter mission (EVEEJ) using SCEA.

Mission Parameter	Zero-DSM Model	MGADSM Model
Departure date	09-Mar-2020 02:32:15	17-Apr-2020 19:13:36
Departure impulse, km/s	3.4726	3.3236
Venus gravitational assist date	07-Jul-2020 11:34:05	04-Oct-2020 18:13:50
Earth gravitational assist date	21-Nov-2022 22:15:29	11-Aug-2022 17:30:55
DSM date	–	05-Sep-2023 09:20:1
DSM impulse, km/s	–	0.1482
Earth gravitational assist date	02-Aug-2024 19:24:44	05-Aug-2024 06:51:12
Arrival date	16-Apr-2027 15:23:34	24-May-2027 23:34:02
Arrival impulse, km/s	5.4852	5.4596
Time of flight between gravitational assist maneuvers (days)	120.4/867.4/612.0/ 986.8	170.0/676.0/724.6/1022.7
Total Time of flight (days)	2594	2593
Mission Cost (km/s)	**9.0373**	**8.9314**

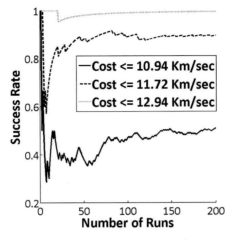

(a) Success rates for three cost values

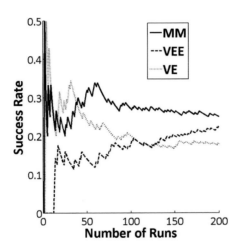

(b) Success rates for three gravitational assist sequences

Figure 7.11: Statistical analysis for SCGA in JEO mission.

implementation. The success rates are shown for several solutions. **Figure** 7.11(a) shows the success rates for finding three different solutions with three different costs. **Figure** 7.11(b) shows the success rates for three different gravitational assist sequences. The sequence (MM) has higher rate, however it is of higher cost. **Figures** 7.12(a) and 7.12(b) show the statistical analysis for the SCDE implementation. **Figure** 7.12(a) shows the success rates for three different solutions, and **Fig.** 7.12(b) shows the success rates in finding three different gravitational assist sequences. Different settings are used in generating **Figs.** 7.12(a)

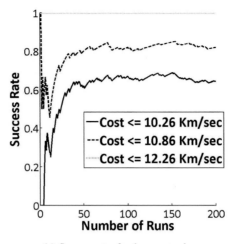

(a) Success rates for three cost values

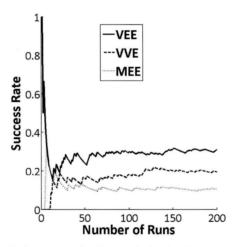

(b) Success rates for three gravitational assist sequences

Figure 7.12: Statistical analysis for SCDE in JEO mission.

and 7.12(b). Both the SCDE and the SCGA find the sequence (VEE) with about the same success rates. In the JEO mission, the success rate of the SCGA is higher than the success rate of SCDE in finding the best known cost.

7.3   Comparisons and Discussion

In all the test cases presented in this chapter, the SCEA method demonstrated the new search capabilities in optimizing an objective function of variable-size design space. For the Earth-Mars mission, SCEA found the known optimal planet sequence (EVM) and the DSM structure. Both the hidden genes genetic algorithms and the dynamic-size multiple population genetic algorithm reported the same solution presented in this chapter. For the JEO mission, the SCEA automatically found the gravitational assist sequence (EVEEJ) and the DSM structure. The obtained solution has a cost of 8.93 km/sec. The hidden genes genetic algorithms found another solution with the same gravitational assist sequence (EVEEJ) but with one DSM in the first leg and a cost of 8.988 km/sec [1]. Reference [1] found another solution with a gravitational assist sequence of (EEEJ), three DSMs, and a cost of 8.92 km/sec.

For the Cassini 2-like mission, the obtained solution using SCEA has a swing-by sequence of (EVVEJS) and a cost of 9.4 km/sec. The best known solution in the literature has the same swing-by sequence, cost of about 8.4 km/sec [79, 90, 115, 67, 49]. It is important to note that the SCEA searches for the optimal sequence of swing-bys autonomously, which means that the SCEA is attempting to solve a more complex problem. In addition, assuming that the optimal sequence is found, it is possible to tune the obtained solution to further reduce the cost through holding the swing-by sequence fixed and using a local optimization tool to tune the rest of the variables. The same concept was also implemented with the hidden genes genetic algorithms. The main advantage of the SCEA is the demonstrated capability to search for the optimal solution in this variable-size design space, and obtaining the correct sequence of swing-bys.

This chapter presented the structured chromosome concept on two optimization algorithms: the differential evolution and genetic algorithms. In all results the SCDE outperforms the SCGA. This is not surprising since other referencing has reported the same conclusion for this type of trajectory optimization problem such as references [2]. Extensive tests have been conducted to determine the optimization algorithm parameters such as the the number of generations, the niching radius, and the black list age. **Figures** 7.13, 7.14, and 7.15 show the effects of these parameters on the obtained solution cost, where p is the population size, gen is the number of generations, age is the number of times a solution is obtained within the specified radius of a local optimal before moving this local optimum to the black listed solutions, and reset is the maximum length of the black list; if the number of black listed solutions reached the maximum, all the back listed solutions will be cleared. The horizontal line inside the box represents the median cost obtained. The boundaries of the box contain 50% of the obtained solutions, 25% of the obtained solutions are above the box upper bound, and 25% of the obtained solutions are below the box lower bound. The maximum and minimum costs are also shown in the figure.

Figure 7.13 shows how the cost changes with various numbers of generations for the Cassini 2-like mission. The results in **Fig.** 7.13 is intuitive, the higher number of generations the better is the solution. **Figure** 7.14 shows the results of two experiments with two different niching radii. In **Fig.** 7.14, for a population

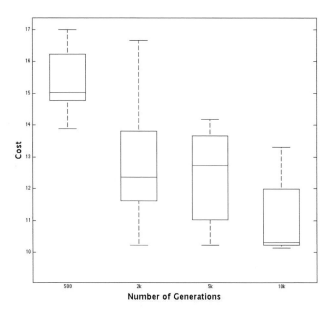

Figure 7.13: Effect of changing the number of generations on the obtained solution for Cassini 2-like mission, population size is 500 and cost is in km/sec.

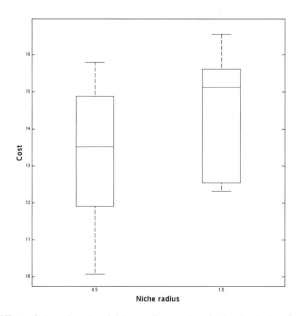

Figure 7.14: Effect of changing the niching radius on the obtained solution for Cassini 2-like mission, population size is 500,no. of generations is 5000, age is 10,000, reset is 1000, and cost is in km/sec.

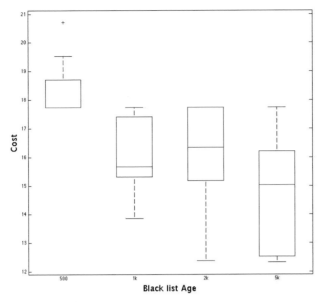

Figure 7.15: Effect of changing the black list age on the obtained solution for Cassini 2-like mission, population size is 500,no. of generations is 2000, rad is 5, reset is 1000, and cost is in km/sec.

of 500 members, number of generations of $5k$ ($= 5000$), age $= 10000$, and a reset $= 1000$, each experiment was run 10 times and the results of each ten runs are presented in the box. The results show that a smaller niching radius is better. **Figure** 7.15 shows that using a higher age results in better solutions.

Chapter 8

Dynamic-Size Multiple Population Genetic Algorithms

This chapter presents a genetic-based intuitive method developed to handle global VSDS optimization problems. Subpopulations, each of them is of fixed-size design spaces, are randomly initialized. Standard genetic operations are carried out for a stage of generations. A new population is then created by reproduction from all members in all subpopulations based on their relative fitnesses. The resulting subpopulations have different sizes from their initial sizes in general. The process repeats, leading to an increase in the size of subpopulations of more fit solutions and a decrease in the size of subpopulations of less fit solutions. This method is called Dynamic-Size Multiple Population Genetic Algorithms (DSMPGA). In space trajectory optimization, for instance, this method has the capability to determine the number of swing-bys, the planets to swing by, launch and arrival dates, and the number of deep space maneuvers as well as their locations, magnitudes, and directions in an optimal sense. This chapter will present the method with several examples on its implementation to the interplanetary trajectory optimization problem. The method can be implemented in other applications as well. The DSMPGA was first published in reference [8].

8.1 The Concept of DSMPGA

Genetic operations, such as crossover, are defined only for fixed-length string populations. Many engineering optimization problems, including the MGADSM problem defined in **Section** 2.2, pose variable-length strings. To overcome this problem, the concept of dynamic-size multiple population genetic algorithms (DSMPGA) is introduced.

The concept of the DSMPGA is to create an initial population that consists of subpopulations, as shown in **Fig.** 8.1. All subpopulations are processed in parallel. All the members in each subpopulation will have the same chromosome length. Different subpopulations will have different chromosome lengths. The standard genetic operations will be applied to each subpopulation for a certain number of generations (a stage of generations), as shown in **Fig.** 8.1. The fitness of all members in the overall population is evaluated at the end of each stage. The selection operation is then carried out across the board (from all subpopulation), based on members fitnesses, regardless of the chromosome length. The more fit

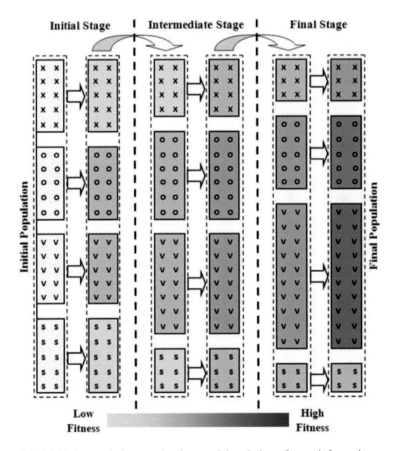

Figure 8.1: Multiple populations evolve in parralel and share fitness information at every stage. The size of each subpopulation in a stage depends on its average fitness in the previous stage relative to the other subpopulations.

members from all subpopulations will be reproduced to subsequent stages in their respective subpopulations. Hence, the size of each subpopulation will change from one stage to another. Subpopulations that have higher (or lower) fitness members will increase (or decrease) in size. Subpopulations are then processed for another stage of generations; a selection at the end of the new stage is then carried out based on the memebers' fitnesses, and so on.

This approach will enable the use of standard genetic operations in VSDS problems. **Figure** 8.1 shows an illustration for this concept. Each symbol represents solutions of the same chromosome length. Subpopulations, each including the same symbol, are stacked in one big population. Subpopulations will change their sizes from one stage to another based on the members' fitnesses, leading to the evolution of more fit members.

There are two options for the sizes of the initial subpopulations (number of individuals in each subpopulation.) The first option is to start with equal sizes for all subpopulations. The second option, which is adapted in this method for the MGADSM problem, is to start with an initial size for each subpopulation that is proportional to the number of design variables in that subpopulation. This choice is selected based on the fact that a problem with a higher number of design variables needs a bigger size population for better convergence. The proportionality constant is c. The size of each subpopulation, then, varies in subsequent generations. The overall population size, however, remains constant and is equal to the summation of all initial subpopulation sizes.

At the end of each stage, an evaluation process is performed to resize all subpopulations based on the relative fitnesses of all individuals in all subpopulations. A roulette selection method is applied to choose parents for the next stage from the overall population of the current stage. The roulette method simulates a roulette wheel, with the area of each segment proportional to its fitness. A higher fitness individual has a bigger segment area. A random number is then used to select one of the sections with a probability equal to its area. The selected individuals are separated according to their associated subpopulation.

To guarantee that the fittest individuals of each subpopulation survive to the next stage, a small portion of the fittest individuals, elite count, is specified. The elite count is selected to be 10% in each subpopulation. These elite individuals guarantee that each subpopulation will be presented in the whole population until the final stage with at least a small size subpopulation. This step also allows a subpopulation with a small size to recover back to a bigger size if more fit members are generated in subsequent generations.

8.2 Application: Space Trajectory Optimization

The DSMPGA is implemented to solve the MGADSM in this section. The mathematical formulation of the problem is presented in **Section** 2.2. The independent design variables to be optimized are also described in **Section** 2.2. The objective is set to minimize the total cost, Δv, of the trajectory for a MGADSM mission that consists of m gravity assist maneuvers and n deep space maneuvers. Equation (2.24) shows the total cost of the mission and Eq. (2.25) shows the fitness function. **Section** 9.1 also presents a brief description for the continuous and discrete optimization variables in this problem. It is recommended to review **Section** 2.2 and **Section** 9.1 before proceeding in this section.

For the MGADSM problem, the total number of independent design variables depends on the number of swing-by maneuvers, m, and the number of DSMs in every leg, $n_1, \cdots, n_{i+1}$. Selecting different values for these discrete design variables changes the mission scenario and the DSM structure of the whole trajectory. Thus, changing the mission scenario and/or the DSM structure is accompanied by a change in the length of the chromosome. This type of problem cannot be handled with the standard genetic algorithm because of the variation of string lengths within the same population. Thus, the DSMPGA method is implemented. For the MGADSM problem, each subpopulation represents a single DSM structure with a fixed-length mission scenario. This means that, for each subpopulation, the number of DSMs in each leg as well as the number of swing-by maneuvers are constants. The chromosome length, L_s, of a specific subpopulation is calculated as follows:

$$L_s = 3 + 2m_s + j + 2(m_s - z_s) + 3k_s \qquad (8.1)$$

where z is the number of swing-by maneuvers which are followed by a zero-DSM leg. The subscript s represents the sub-population. The whole population consists of a set of sub-populations, and each sub-populations has a fixed chromosome length. The number of sub-populations depends on the lower and upper bounds of the key design variables (m, n_l). For a certain value of m, all possible combinations of DSM structure are taken into account, as shown in **Fig.** 8.2. For example, if $m = 1$, then the trajectory consists of two legs. Consider the lower and upper bounds of n_l for each leg are zero and one, respectively. Therefore, there are four possible combinations of the DSM structure. The four sub-populations in this case are: zero-DSM in both legs, a single DSM in the first leg with zero-DSM in the second leg, a single DSM in the second leg with zero-DSM in the first leg, and a single DSM in both legs. These four sub-populations may have different chromosome length.

The independent design variables m and n_l are not presented in the chromosome. The bounds of these two variables determines the problem size by specifying the number of sub-populations. Each sub-population represents a single value of m and a single combination of n_l. The initial population size, number of individuals, of each sub-population is selected to be proportional with the number of the associated design variables. This criteria is applied based on the fact of higher number of design variables needs a bigger population size. The proportional constant is defined as the initial population constant c. The initial whole population is the summation of all initial sub-populations. Standard genetic al-

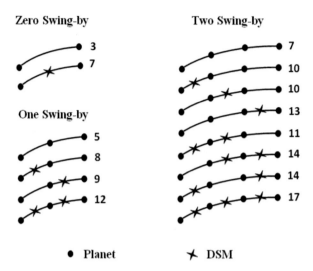

Figure 8.2: DSM structure for different scenarios, the number of independent design variables are presented for each sub-population.

gorithm is separately performed for each sub-population with equal number of generations. The optimization problem is performed over consequent stages. The output from the first stage is the input of the next stage, and so on. The population size is fixed over the consequent stages. But the sub-population size is changing according to the performance of the fitness function. The most fit sub-population is expected to increase in size, while the least fit sub-population is decreased in population size.

After each stage, an evaluation process is performed to arrange the whole population individuals based on the fitness function. To guarantee that the fittest individuals of each sub-population are survived to the next stage, a small number of the fittest individuals, elite count, is specified. The common value of the elite count is 10% of each sub-population. These elite individuals guarantee that each sub-population will be presented in the whole population till the final stage with at least a small number of population. This step is used because in some cases the fitness performance improves with more run generations. A roulette selection method is then applied to choose parents for the next stage based on their scaled values from the fitness scaling function. Roulette simulates a roulette wheel with the area of each segment proportional to its fitness. A higher fitness individual has a bigger segment area. A random number is then used to select one of the sections with a probability equal to its area. The selected individuals are separated according to their associated sub-population. The resulting sub-population is considered as an initial population in the next stage. The roulette selection guarantees that the sub-population with a higher fitness individuals will be increased in size.

This DSMPGA is tested for several interplanetary missions ranging from simple to complex missions. The optimization algorithm is designed to find the mission scenario (the number of swing-bys and the planets to swing by) as well as the rest of the independent design variables: the times of swing-by, the number of DSMs, the times of DSMs, the magnitudes/directions of DSMs, and the departure/arrival dates. The size of the design space is controlled by the bounds of the independent design variables. The minimum and maximum number of the DSMs in the whole trajectory are important parameters which specify the number of sub-populations. In some missions, one of the objectives is to minimize the number of applied DSMs as well as the total mission cost. The number of revolutions in each leg is also optimized by the DSMPGA tool. The time of flight (TOF) of each leg and the time of applying each DSM are selected as design variables to allow for multi-revolution transfers. Lambert's problem is solved once in each leg. Based on the position vectors and the TOF, Lambert's solution may have both single and multi-revolution transfers. The selection criterion is as follows: the selected lambert's transfer should minimize the former maneuver cost. This maneuver could be a departure impulse, a DSM, or a swing-by maneuver. In the final leg, the selected lambert's transfer will minimize the former maneuver cost plus the arrival impulse cost.

The main operations in the standard GA are coding, evaluation, selection, crossover and mutation. GA tuning parameters depend on the specific problem. A scattered crossover operation is conducted with a probability varied from 0.8 to 0.9. An adaptive feasible mutation function is considered so that the design variables bounds are satisfied. Rank scaling function is applied. Roulette wheel is used in selection operation. Wider initial ranges are used in order to increase the diversity in the population. The solution obtained by the genetic algorithm is not necessary an optimal solution, nor at a local minimum. Therefore, a constrained nonlinear optimization technique is used to improve the solution by finding the closest local minimum to that solution. The local optimizer is only optimizing over the continuous design variables, not the discrete variables. The genetic algorithm fittest solution is used as an initial guess in the local search algorithm.

8.3   Numerical Examples

This section presents a number of numerical examples for interplanetary space missions trajectory optimization. Comparisons to other solutions in the literature is presented to validate the obtained results.

Example 8.1. Earth-Mars Mission

A MGADSM trajectory is optimized for the Earth-Mars mission (EM). The lower and upper bounds, for all design variables, are listed in **Table** 8.1. The

Table 8.1: Bounds of design variables for Earth-Mars mission.

Design Variables	Lower Bound	Upper Bound
No. of swing-by maneuvers, m	0	2
Swing-by planets identification numbers, $P_1,..., P_i$	1 (Mercury)	8 (Neptune)
No. of DSMs in each mission's leg, $n_1,..., n_{i+1}$	0	1
Flight direction, f	Posigrade	Retrograde
Departure date, t_d	01-Jun-2004	01-Jul-2004
Arrival date, t_a	01-Apr-2005	01-Jul-2005
Time of flight (days)/leg, $T_1,..., T_i$	40	300
Swing-by normalized pericenter altitude, $\bar{h}_1,..., \bar{h}_i$	0.1	10
Swing-by plane rotation angle (rad), $\eta_1,..., \eta_i$	0	2π
Epoch of DSM, $\varepsilon_1,..., \varepsilon_j$	0.1	0.9
DSM (km/s), $\Delta v_1,..., \Delta v_k$	-5	5

maximum possible number of swing-by maneuvers is selected to be two. The number of DSMs in each leg is selected to be either zero or one impulse. As can be seen from **Table** 8.1, a gravity assist maneuver could be performed with any planet in the solar system, from Mercury to Neptune. The direction of flight can be posigrade or retrograde. The time of flight of each leg, except the last one, is selected between 40 and 300 days.

The population is divided into 14 sub-populations (2 for zero swing-by, 4 for one swing-by, and 8 for two swing-by maneuvers). The maximum number of design variables is 17 which is used in the last sub-population (two swing-by with a single DSM in each leg). The other sub-populations have different numbers and combinations of the independent design variables. The initial population constant c is selected to be 10. The whole population size is 1400. The number of generations in each standard GA stage is 30, and there are 5 stages. 10% is selected as elite count with a minimum of 10 individuals in each sub-population. A local optimizer uses the fittest GA solution as an initial guess to find a local minimum. The resulting solution has a single swing-by maneuver at Venus, with a total cost of 10.703 km/s. The fittest trajectory has a single DSM in each leg as shown in **Table** 8.2. **Figure** 8.3 shows the obtained EVM trajectory.

Example 8.2. Earth-Jupiter Mission

The DSMPGA tool is used to investigate the Earth-Jupiter mission. The objective is to design an optimal MGADSM trajectory with minimum propellent consumption. Wide ranges of all design variables are allowed in the optimization to explore more of the design space. The design variables' bounds are listed in **Table** 8.3. The population is divided into 14 sub-populations. The maximum number of design variables is 17 which is used in the last sub-population (two swing-by with a single DSM in each leg). The other sub-populations have different numbers and combinations of the independent design variables. The initial

Table 8.2: MGADSM solution trajectory for the EVM mission.

Mission Parameter	MGADSM Scenario
Departure date,t_d	01-Jun-2004 00:00:00
Departure impulse (km/s)	4.386
DSM date	02-Sep-2004 17:06:11
DSM impulse (km/s)	0.296
Venus Swing-by date	17-Nov-2004 13:51:17
Pericenter altitude (km)	7701.2
DSM date	30-Apr-2005 18:11:07
DSM impulse (km/s)	0.626
Arrival date	19-May-2005 01:47:07
Arrival impulse (km/s)	5.392
Time of flight (days)	169.58 , 182.49
Mission duration (days)	352.07
Mission Cost (km/s)	**10.7**

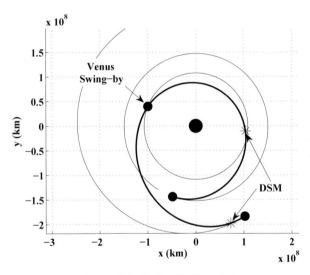

Figure 8.3: Optimal EVM mission.

population constant c is selected to be 20. The whole population size is 2800. The number of generations in each standard GA stage is 40, and there are 5 stages. 10% is selected as elite count with a minimum of 10 individuals in each sub-population. A local optimizer uses the fittest GA solution as an initial guess to find a local minimum.

Table 8.3: Bounds of Earth-Jupiter mission's design variables.

Design Variables	Lower Bound	Upper Bound
No. of swing-by maneuvers, m	0	2
Swing-by planets identification numbers, $P_1,..., P_i$	1 (Mercury)	8 (Neptune)
No. of DSMs in each mission's leg, $n_1,..., n_{i+1}$	0	1
Flight direction, f	Posigrade	Retrograde
Departure date, t_d	01-Sep-2016	30-Sep-2016
Arrival date, t_a	01-Sep-2021	31-Dec-2021
Time of flight (days)/leg, $T_1,..., T_i$	80	800
Swing-by normalized pericenter altitude, $\overline{h}_1,..., \overline{h}_i$	0.1	10
Swing-by plane rotation angle (rad), $\eta_1,..., \eta_i$	0	2π
Epoch of DSM, $\varepsilon_1,..., \varepsilon_j$	0.1	0.9
DSM (km/s), $\Delta v_1,..., \Delta v_k$	-5	5

The resulting fittest scenario is a two swing-by trajectory with swing-bys around Venus then Earth (EVEJ). The solution is a posigrade multi-revolution trajectory. The obtained trajectory has a single DSM in the second leg (VE) with a total cost of 10.125 km/s, as shown in **Table** 8.4. The optimal EVEJ trajectory is shown in **Fig.** 8.4. The DSM amplitude is 2.14 m/s which seems an insignificant value with respect to the total trajectory cost. It is expected that this small DSM might be vanished with more iterations. A powered swing-by maneuver is obtained with the second swing-by planet. The post-swing-by impulse is 0.443 km/s applied during the Earth swing-by maneuver.

Table 8.4: MGADSM solution trajectory for the EVEJ mission.

Mission Parameter	MGADSM Scenario
Departure date, t_d	01-Sep-2016 01:54:21
Departure impulse (km/s)	3.487
Venus Swing-by date	05-Sep-2017 10:21:59
Pericenter altitude (km)	1339.21
DSM date	25-May-2018 09:03:27
DSM impulse (km/s)	0.00214
Earth Swing-by date	30-Mar-2019 06:01:26
Post-swing-by impulse (km/s)	0.443
Pericenter altitude (km)	637.8
Arrival date	29-Sep-2021 00:31:15
Arrival impulse (km/s)	6.193
Time of flight (days)	369.35 , 570.82 , 913.77
Mission duration (days)	1853.94
Mission Cost (km/s)	**10.125**

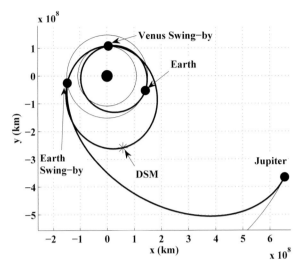

Figure 8.4: Optimal EVEJ mission.

Example 8.3. Messenger (easy version)

Messenger is the first mission to explore the planet Mercury. Messenger trajectory follows a path through the inner solar system to rendezvous Mercury. The easy version is considered to simplify the problem structure by excluding the resonant swing-bys around planet Mercury [30]. The DSMPGA tool is used to design an optimal MGADSM trajectory of the rendezvous mission to Mercury. The design variables' bounds are listed in **Table** 8.5. These values are selected to be consistent with the problem database listed in reference [30].

There are 28 sub-populations with different DSM structure. The maximum number of DSMs is four while the minimum number of DSMs is zero. The maximum number of design variables is 22 which is used in the last sub-population (three swing-by with a single DSM in each leg). The other sub-populations have different numbers and combinations of the independent design variables. The initial population constant c is selected to be 15. The whole population size is 5670. The number of generations in each standard GA stage is 40, and there are 5 stages. 10% is selected as elite count with a minimum of 10 individuals in each sub-population. A local optimizer uses the fittest GA solution as an initial guess to find a local minimum.

The obtained optimal solution is a three swing-by trajectory with the same planet sequence as of the actual Messenger mission (easy version) scenario (EEVVY). The solution is a posigrade trajectory with three DSMs, a single DSM in each leg except the second leg. The obtained solution is listed as the first scenario in **Table** 8.6. The first scenario solution has a total cost of 8.6312 km/s, and is shown in **Fig.** 8.5. To improve the mission cost, wider bounds are used for the

Table 8.5: Bounds of Messenger mission's design variables.

Design Variables	Lower Bound	Upper Bound
No. of swing-by maneuvers, m	1	3
Swing-by planets identification numbers, $P_1,..., P_i$	2 (Venus)	
No. of DSMs in each mission's leg, $n_1,..., n_{i+1}$	0	4 (Mars) 1
Flight direction, f	Posigrade	Retrograde
Departure date, t_d	01-Jan-2003	31-Mar-2003
Arrival date, t_a	01-Jan-2006	30-Jun-2006
Time of flight (days)/leg, $T_1,..., T_i$	30	400
Swing-by normalized pericenter altitude, $\bar{h}_1,..., \bar{h}_i$	0.1	5
Swing-by plane rotation angle (rad), $\eta_1,..., \eta_i$	0	2π
Epoch of DSM, $\varepsilon_1,..., \varepsilon_j$	0.1	0.9
DSM (km/s), $\Delta v_1,..., \Delta v_k$	-5	5

Table 8.6: Optimal MGADSM trajectory of Messenger mission (easy version) - First Scenario.

Mission Parameter	First Scenario
Departure date	17-Mar-2003 00:02:43
Departure impulse (km/s)	1.421
DSM date	20-Jun-2003 08:04:26
DSM impulse (km/s)	0.904
Earth Swing-by date	20-Apr-2004 00:02:43
Post-swing-by impulse (km/s)	0.00011
Pericenter altitude (km)	4749.47
DSM date	-
DSM impulse (km/s)	-
Venus Swing-by date	15-Oct-2004 22:58:49
Pericenter altitude (km)	12251.66
DSM date	20-Jun-2005 02:59:57
DSM impulse (km/s)	0.258
Venus Swing-by date	11-Aug-2005 04:39:48
Pericenter altitude (km)	605.2
DSM date	05-Oct-2005 22:29:27
DSM impulse (km/s)	1.448
Arrival date	08-Feb-2006 03:20:21
Arrival impulse (km/s)	4.6
Time of flight (days)	400, 178.96, 299.24, 180.94
Mission duration (days)	1059.14
Mission Cost (km/s)	**8.6312**

time of flight design variables. The upper bound of the time of flight is changed to 500 days instead of 400 days. Only 8 sub-populations are considered in the second iteration. These sub-populations represent only three swing-by scenarios with a single DSM in the first leg. The other legs could have zero or one DSM. The initial population constant c is selected to be 30. The whole population size is 4200. The number of generations in each standard GA stage is 40, and there are 5 stages. The obtained solution is listed in **Table** 8.7 as the second scenario. It has the same planet sequence of the first scenario but with a different DSM structure. The optimal trajectory has four DSMs, a single DSM in each leg, with a total cost of 8.203 km/s. The second scenario trajectory is shown in **Fig.** 8.6. In the second scenario, the DSM amplitude in the second leg (VV) is 0.24 m/s which seems an insignificant value with respect to the total trajectory cost. It is expected that this small DSM might be vanished with more iterations.

Table 8.7: Optimal MGADSM trajectory of Messenger mission (easy version) - Second Scenario.

Mission Parameter	Second Scenario
Departure date	14-Feb-2003 08:13:21
Departure impulse (km/s)	1.002
DSM date	05-Jul-2003 10:56:08
DSM impulse (km/s)	0.906
Earth Swing-by date	22-Apr-2004 10:36:27
Post-swing-by impulse (km/s)	-
Pericenter altitude (km)	4155.028
DSM date	27-May-2004 00:20:53
DSM impulse (km/s)	0.00024
Venus Swing-by date	18-Oct-2004 07:47:17
Pericenter altitude (km)	11395.952
DSM date	04-Mar-2005 05:01:11
DSM impulse (km/s)	0.192
Venus Swing-by date	13-Aug-2005 15:36:22
Pericenter altitude (km)	605.2
DSM date	07-Oct-2005 07:00:37
DSM impulse (km/s)	1.514
Arrival date	08-Feb-2006 15:46:58
Arrival impulse (km/s)	4.589
Time of flight (days)	433.1, 178.88, 299.33, 179.01
Mission duration (days)	1090.32
Mission Cost (km/s)	**8.203**

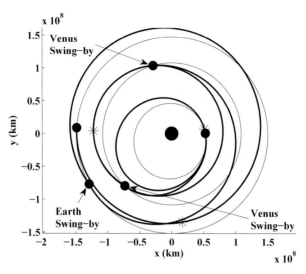

Figure 8.5: Optimal Messenger mission (easy version), first scenario (3 DSMs).

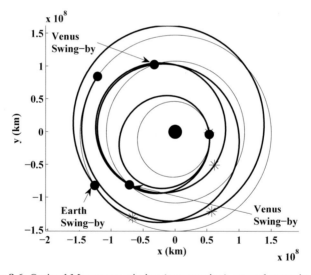

Figure 8.6: Optimal Messenger mission (easy version), second scenario (4 DSMs).

8.4 Discussion

The Earth-Venus-Mars mission trajectory optimization has been presented in the literature [91]. The extended primer vector theory is used to obtain the solution presented in reference [91]. The trajectory has a single swing-by maneuver and a single DSM. In implementing the primer vector method, fixed departure and arrival dates were assumed with 340 days mission duration. The Venus swing-by time was also constrained to occur at 165 days from departure. The resulting solution has a DSM of 68.7 m/s in the first leg at 96.08 days from mission start date. The total cost of the mission is 10.786 km/s [91]. This same problem has been solved by the author using the ITO-HGGA tool (interplanetary trajectory optimization using the hidden genes genetic algorithms) [49]. The total cost of the mission is 10.728 km/s. The trajectory has a single Venus swing-by and a single DSM (180.1 m/s) in the first leg. The solution obtained using the DSMPGA tool in this paper has one swing by Venus and one DSM in each leg, 0.296 and 0.626 km/s, respectively. The total cost of the mission is 10.7 km/s, as shown in **Table** 8.2. The reduction in the total cost obtained using the DSMPGA tool, as compared to that of references [91] and [49], is accompanied by freely chooses the mission's departure, swing-by, and arrival dates, without significantly changing the total mission duration.

A minimum-cost solution trajectory for the Earth-Jupiter mission is addressed in reference [91]. A fixed planet sequence EVEJ is assumed. The departure, arrival, and swing-by dates were also assumed fixed, with a launch in 2016 and a mission duration of 1862 days. The primer vector theorem solution has four DSMs. Two DSMs are applied in the first two legs. The total transfer cost for this solution is 10.267 km/s. The DSMPGA tool developed in this paper is able to find automatically the known swing-by sequence, EVEJ, for the Jupiter mission in 2016. As listed in **Table** 8.4, the optimal MGADSM solution has 10.125 km/s total cost with a single DSM in the whole trajectory. The ITO-HGGA tool was used to solve the same mission in reference [49]. The obtained trajectory from ITO-HGGA has the same planet sequence and DSM structure but with higher total cost, 10.178 km/s [49].

The Messenger mission (easy version) trajectory design problem has been addressed in several studies [30, 68], where it is always assumed that a fixed swing-by sequence (EEVVY) is known. According to the GTOP database, the best solution is found by F. Biscani, M. Rucinski and D.Izzo, using PaGMO, a new version of DiGMO based on the asynchronous island model [30]. The total cost is 8.63 km/s with a single DSM in each leg. Olympio and Izzo [68] recently developed an algorithm to find only the optimal DSM structure in a given trajectory scenario. They used the same ephemeris tool of the GTOP database [30] but with a free number of applied DSMs. For the Messenger planets' sequence described in the GTOP database [30], reference [68] found a trajectory with a total cost of 8.494 km/s. The DSM structure is 2-0-1-1 in consequent leg respectively. The ITO-HGGA tool failed to find an optimal solution for the Messenger mission [49]. The developed tool in this paper, DSMPGA, is used twice to investigate this mission. In the first iteration, the same design variables' bounds of the GTOP database are used. The obtained optimal solution has a total cost of 8.6312 km/s (as seen in **Table** 8.6), which is very close to the reported result in the GTOP database [30]. The DSMPGA tool has the advantage of finding the planets sequence as well as the DSM structure. In the second iteration, a better solution is found with a total cost of 8.203 km/s (as seen in **Table** 8.7.) The significant improvement in fuel consumption is a result of using wider time of flight windows.

Figure 8.7 shows the change in each sub-population size (number of individuals) with the consequent stages. Eight sub-populations are shown which represent the second iteration of the Messenger easy mission. The initial population of each sub-population has different number of individuals depend on the number of independent design variables in each set. With consequent stages, the size of each sub-population varies according to the behavior of the fitness function. The higher fit sub-populations are increased in size, while the lower fit sub-populations are decreased in population size.

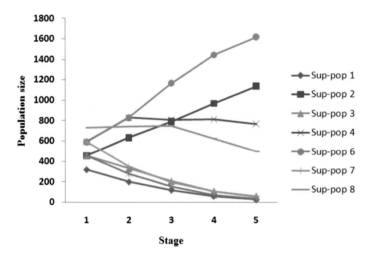

Figure 8.7: Sub-populations size performance for Messenger mission (easy version), second iteration.

APPLICATIONS   IV

Chapter 9

Space Trajectory Optimization

9.1   Background

The optimization of interplanetary trajectories continues to receive a great deal of interest [119]. In interplanetary missions, it is usually desired to send a spacecraft to rendezvous with a planet, or an astroid. The interplanetary trajectory design problem can be addressed either in a two-body or a three-body dynamics framework. It can be addressed assuming the usage of only impulsive thrusters onboard the spacecraft, or assuming the availability of a continuous thrust. The famous Patched Conic approach for interplanetary trajectory design assumes a two-body dynamics model and the usage of impulsive thrusters only.

In Patched Conic mission design, it is observed that a Deep Space Maneuver (DSM) reduces the cost of a simple two-impulse interplanetary transfer (e.g., the cost of the Earth-Mars mission can be reduced by adding an impulse, almost mid-way, in addition to the initial and final impulses.) For a transfer between two non-coplanar orbits, a DSM reduces the out of plane components of the required delta-v, and thus the total delta-v [84]. The usage of a DSM to reduce the total delta-v has the benefit of reducing the propellant mass, and hence, allows for a larger size payload or a smaller launch vehicle. DSMs could also generate new families of trajectories that would satisfy very specific mission requirements not achievable with ballistic trajectories [15]. Navagh applied the primer vector theory to determine where and when to use a single DSM to reduce the cost of a trajectory [84]. He applied DSMs to a Mars round trip mission to determine their effect on the launch opportunities. He also studied cycler trajectories and Mars mission abort scenarios. Later, the broken-plane maneuver impact on the total mission delta-v, the arrival v-infinity, and the launch energy was investigated, for Earth to Mars trajectories [15]. The extra delta-v was suggested to be applied close to the halfway point of the near 180° transfer [15].

The optimization problem, in its general form, aims to minimize some cost function; usually the overall cost of the mission in terms of the fuel expenditure. Another possibility is to optimize mission trajectory to achieve minimum mission duration. Interplanetary missions usually benefit from the gravity assist maneuvers to reduce the overall cost of the mission. In some cases, applying a deep space impulsive maneuver reduces the overall mission cost. The launch and arrival dates also affect the mission cost. Hence, the space trajectory optimization process determines the optimal values for the following parameters: the number of swing-bys, the planets to swing-by, the number of deep space maneuvers and their amounts and locations, and the departure and arrival dates. This optimization problem, in its general form, is challenging. Several optimization algorithms were developed in the literature.

Izzo et al developed a deterministic search space pruning algorithm to investigate the problem of multiple gravity-assist (MGA) interplanetary trajectories design [65]. The developed tool requires the user to specify the gravity assist sequence (i.e., the number of swing-bys and the planets to swing-by), and the pruning technique locates efficiently all the interesting parts in the search space of the DSMs variables [65].

Olympio and Marmorat [91] studied the global optimization of multi gravity assist trajectories with deep space maneuvers (MGADSM). They implemented a pruning strategy on the search space to find fit trajectories. To that end, a stochastic initialization procedure, combined with a local optimization tool, were used to provide a set of local optimum solutions. The primer vector theory was extended to study the multi gravity assist trajectories. The theory was applied to determine the optimal number of DSMs and their amounts. This technique was verified using several interplanetary missions. This method needs an efficient local optimization algorithm and a good management to find a solution for complex problems [91]. In this method, the user fixes the planets to swing-by before the optimization can take place.

Later, Olympio and Izzo applied the interaction prediction principle to decompose the MGA problem into sub-problems by introducing and relaxing boundary conditions [68]. Thus, parallel sub-problems could then be solved. The developed algorithm was able to efficiently calculate the optimum number of DSM impulses and their locations.

Evolution algorithms were implemented to solve the MGADSM problem. In particular, Genetic Algorithms (GAs) have been used widely in the literature to solve several orbital mechanics problems. GAs have been used to solve the fuel-optimal spacecraft rendezvous problem [71], to design orbits with lower average revisit time over a particular target site [70], to design natural orbits for ground surveillance [9], and to design constellations for zonal coverage [36]. The usage of GAs, indeed, exceeded that to cover topics such as the design of near-optimal low-thrust orbit transfers [97].

Genetic algorithms are optimization techniques, based on the Darwinian principle of the survival of the fittest, which perform a stochastic search of initial conditions that maximize a given objective "Fitness" function [62]. The GAs have global search capability, yet they do not require objective function's derivatives [36]. GAs are particulary efficient for the type of problems where it is not necessarily to find the optimal solution, but rather, few reasonably good solutions.

A formulation for the N-impulse orbit transfer that integrates evolutionary algorithms and a Lambert's problem formalism results in a more efficient search algorithm [10, 115]. This integration of Lambert's problem results in reducing significantly the size of the design space to include only those members who satisfy Lambert's problem solution. In that context, the GAs were implemented to obtain the optimum N-impulses for non-coplanar elliptical interplanetary orbit transfers [10, 50].

Vasile and Pascale used an evolutionary-based algorithm with a systematic branching strategy to optimize the problem of MGADSM [115]. They developed a design tool (IMAGO) that allowed for more exploration of the solution domain through balancing the local convergence and the global search. The search space was reduced by performing a deterministic step at every new run. For this type of solution algorithms, it is usually needed to run the program several times on each specific problem to provide reliable results [115]. IMAGO was able to calculate many typical optimal sequences to Jupiter mission. For the complex trajectories, Cassini and Rosseta [115], IMAGO was used assuming a fixed planet sequence and wide ranges of design variables, based on the problem knowledge.

As can be seen from the previous discussion, often times there is the need that the mission designer determines the number of swing-bys and selects the planets to swing-by. The motivation of this study is to develop an optimization algorithm that can compute the number of swing-bys and the planets to swing-by, along with the rest of the classical MGADSM design variables, in an attempt to automate the design process. The objective of this study is to develop an optimization algorithm and a software tool that have the capabilities to find, in an optimal sense, the values for the following design variables: optimal number of swing-bys, the planets to swing-by, the times of swing-bys, the optimal number of DSMs, the amounts of these DSMs, the times at which these DSMs are applied, the optimal launch and arrival dates, and the optimal flight direction for the mission. The search space of the developed algorithm includes trajectories with muti-revolution trajectories. This is a complex problem characterized by the following: first, some of the design variables are discrete, second, the number of design variables is solution-dependent, and finally the number of design variables becomes rather high in complex missions. The fact that some of the design variables are discrete suggests the use of stochastic optimization techniques such as GAs. Solution-dependent design variables mean that different solutions have different number of design variables. This fact hinders the implementation of standard genetic algorithms in optimization. The mathematical formulation of the problem is presented in **Section** 2.2. The independent design variables to be optimized are also described in **Section** 2.2. In this chapter, the objective is set to minimize the total cost, Δv, of the trajectory for a MGADSM mission that consists of m gravity assist maneuvers and n deep space maneuvers. Equation (2.24) shows the total cost of the mission, and it is written here for convenience:

$$\Delta v = \|\Delta \mathbf{V}_d\| + \sum_{i=1}^{m} \|\Delta \mathbf{V}_{ps_i}\| + \sum_{j=1}^{n} \|\Delta \mathbf{V}_{DSM_j}\| + \|\Delta \mathbf{V}_a\| \qquad (9.1)$$

where $\Delta \mathbf{V}_d$ and $\Delta \mathbf{V}_a$ are the departure and arrival impulses, respectively, $\Delta \mathbf{V}_{ps}$ is the post-swing-by impulse of the powered gravity assist only, and $\Delta \mathbf{V}_{DSM}$ is the applied deep space maneuver impulse. The fitness F_i at a design point is defined as:

$$F_i = \frac{1}{\Delta v} \qquad (9.2)$$

The independent design variables are two categories: discrete and continuous variables, as shown in **Table** 9.1. In **Table** 9.1, i is the maximum possible number of swing-by maneuvers and j is the maximum possible number of total DSMs in the whole trajectory (both i and j are specified by the user). The term DSM is used to define any thrust impulse applied during the mission course except the launch and arrival impulses. The maximum number of independent thrust impulses is k, which can be explained as follows: if x is the maximum number of DSMs in a leg, then there are x independent thrust impulses if this leg is the first one; while for the consequent legs, the number of required independent thrust impulses is $x - 1$. Each thrust impulse $\Delta \mathbf{V}$ consists of three continuous design variables. The pericenter altitudes $\overline{h}$ are normalized with respect to the mean radius of the associated swing-by planet. Using $\overline{h}$ as a design variable limits the resulting pericenter altitudes to feasible values only, when dealing with different swing-by planets with obvious varied radii. The number of DSMs in leg l is D_l. The epoch of a DSM, ε, specifies the time at which the impulsive maneuver is applied, as a fraction of the associated leg transfer time. For most applications, a DSM could be applied from 10% up to 90% of the associated leg time of flight. The flight direction F_d of the whole mission is either retrograde or posigrade.

In the following sections, the methods presented in **Part III** are used to solve this VSDS problem.

Table 9.1: The discrete and continuous independent design variables of the MGADSM problem.

Discrete Design Variables (DDV)	Continuous Design Variables (CDV)
No. of swing-by maneuvers, N_s	Departure date, t_d
Swing-by planets, $P_1,..., P_i$	Arrival date, t_a
Flight direction, F_d	Time of flight, $T_1,..., T_i$
The count of DSMs in each	Pericenter altitudes, $\bar{h}_1,..., \bar{h}_i$
leg, $D_1,..., D_{i+1}$	Rotation angles, $\eta_1,..., \eta_i$
	Epochs of DSMs, $\varepsilon_1,..., \varepsilon_j$
	Thrust impulses, $\Delta v_1,..., \Delta v_k$

9.2 A Simple Implementation of HGGA

In this section, the simple version of HGGA presented in **Chapter** 6 is used to solve the trajectory design VSDS optimization problem. The results of implementing this version of the HGGAs to multiple case studies are presented in this section. This implementation of the simple HGGA to this problem was first published in [49].

9.2.1 Optimization

Given that the maximum possible number of swing-by maneuvers is i, then we create i discrete design variables P_1, P_2, ..., P_i. Each variable P_l determines the planet about which the lth swing-by occurs, as shown in **Fig.** 9.1. The range of the discrete variable P_l is 1 through 8, which are the codes for the planets in the solar system, starting from Mercury to Neptune. The order of the swing-bys is the same as the order of the variables P_l, as shown in **Fig.** 9.1. If the selected number of swing-bys is $N_s < i$, then the first N_s variables $(P_1, P_2, ..., P_{N_s})$ will be selected.

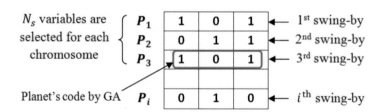

Figure 9.1: Coding of swing-by planets.

For a MGADSM problem, the total number of independent design variables depends on the number of swing-by maneuvers, N_s, and the number of DSMs in every leg, $D_1, ..., D_{i+1}$. Selecting different values for the discrete design variables $N_s, D_1, ..., D_{i+1}$, changes the mission scenario and the number of DSMs in each leg. So, for a given solution, the values of these variables dictate the length of a certain portion in the chromosome of that solution. For example, if $N_s = 2$, then we need to allocate two swing-by portions in the chromosome; however, if $N_s = 3$, then three postions will be needed, for swing-bys, in the chromosome. Thus, changing the mission scenario and/or the number of DSMs is accompanied by a change in the length of the chromosome. Standard genetic algorithms cannot handle this problem because of the variation of chromosomes lengths among different solutions. Genetic operations, such as crossover, are defined only for fixed-length chromosome populations. To overcome this problem, the concept of hidden genes is introduced.

Let L_{max} be the length of the longest possible chromosome (this chromosome corresponds to a trajectory in which the spacecraft performs the maximum possible number of swing-bys and applies the maximum number of DSMs). All chromosomes in the population are allocated a fixed length equal to L_{max}, which is computed, based on the design variables stated in **Table** 9.1, as in Eq. (9.3).

$$L_{max} = 5 + 3i + j + 2(i - q) + 3k \qquad (9.3)$$

where q is the number of swing-by maneuvers which are followed by a no-impulse leg. In a general solution (a point in the design space), some of the variables in the chromosome will be ineffective in cost function evaluation; the genes describing these variables are here referred to as hidden genes. The hidden genes, however, will be used in the genetic operations in generating future generations. **Figure** 9.2 shows typical solution chromosomes where the variables at the top of the figure are the independent design variables.

Because all chromosomes have the same length, standard definitions of genetic algorithms operations can still be applied to this problem. Hidden genes will take part in all genetic operations like normal genes. To illustrate this concept, an example on the crossover operation, between two parent solutions to

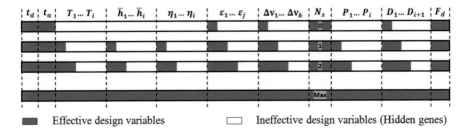

Figure 9.2: Typical chromosomes for the trajectory optimization problem.

generate two children (new) solutions, is here presented. **Figure** 9.3 shows two parent chromosomes and the resulting chromosomes after crossover, where a single point crossover is performed. Consider, for example, the genes representing the number of swing-by maneuvers and the genes representing the numbers of DSMs. The genes representing the number of swing-by maneuvers and the numbers of DSMs are always effective genes. The chromosome length allows for up to 5 swing-bys and up to 2 DSMs in each leg. In the first parent, the genes representing the first two swing-bys are effective genes, while there are hidden genes representing the other three swing-bys. Similarly, the genes representing the numbers of DSMs of legs 1 through 3 ($D_1 : D_3$) are effective genes, while there are hidden genes representing the rest of DSMs variables. In the second parent, the genes representing the first three swing-bys are effective genes, while there are hidden genes representing the other two swing-bys. The genes representing the first 4 DSM's variables are effective genes ($D_1 : D_4$), while there are hidden genes representing the rest of DSMs variables. The crossover operation is carried out as defined in the standard genetic algorithm. The resulting two children are shown in **Fig.** 9.3. The first child has effective genes representing a single swing-by and two DSM variables (The first DSM variable has a single impulse, while the second has a no-impulse trajectory). The effective genes in the second child correspond to a zero-swing-by maneuver with a single DSM maneuver. The rest of the chromosomes of the children are hidden genes which represent the ineffective design variables. Therefore, the generated children could have different scenarios and DSM sequences from those of the origin parents.

This algorithm is tested for several interplanetary missions ranging from simple to complex missions. The optimization algorithm (ITO-HGGA tool) is de-

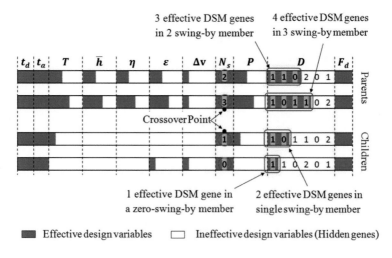

Figure 9.3: Crossover operation.

signed to find the mission scenario (the number of swing-bys and the planets to swing by) as well as the rest of the independent design variables: the times of swing-by, the number of DSMs, the times of DSMs, the amouns/directiosn of DSMs, and the departure/arrival dates. The size of the design space is controlled through the specification of the ranges of the independent design variables.

To reduce the computational cost for complex missions, the problem may be split into two subproblems. A trajectory design optimization can be started by assuming no DSMs in the trajectory (zero DSMs trajectory). This reduces the number of independent design variables by eliminating the following design variables: periapsis altitudes, rotation angles, epoch of DSMs, and thrust impulses. This reduction in the number of design variables allows for exploring wider ranges for each of the remaining design variables. Specifically, we can open the search space for any number of swing-bys with any planets in the solar system; wider ranges of departure and arrival dates and times of flight can be used; and both possible flight directions can be considered. This will result in a set of fit scenarios (zero DSM solutions). For each one of these scenarios, we then allow DSMs to be added to the trajectory and optimize our selection for these DSMs. So, we optimize on the number of DSMs in each leg and their locations/amounts/directions, while maintaining the scenario fixed. The departure/arrival dates and times of flight are allowed to change as design variables, with narrow ranges around the values obtained from the zero DSM solution. This technique of splitting the problem into two sub-problems has the advantage of reducing the computational time needed. This reduction in computational time, comes at a price: part of the design space will not be explored. In general, optimizing a given scenario through adding DSMs, improves the fitness of this trajectory (If a zero DSM trajectory is optimal, then the second step in the optimization process will add no DSMs to the scenario). On the other hand, some solutions are fit only when there are DSMs in the trajectory and the fitness of the corresponding zero DSM scenario is poor (e.g., the MESSENGER mission trajectory is fit when we have a DSM in the first leg and becomes poor if we assume no DSM in the first leg). This means that it is not possible to find the zero DSM solution among the fittest solutions in the first step. For this kind of trajectories, it is not possible to find the optimal solution by splitting the problem into two subproblems.

As an example, consider the Cassini 2 mission trajectory, let the maximum possible number of swing-by maneuvers be four, and let only one impulse, as a maximum, be applied in each leg. In optimizing all the design variables, the required number of independent design variables is 33 (11 DDVs and 22 CDVs). This is relatively a computationally expensive problem, specially if we have wide ranges for the design variables. On the other hand, by solving a zero-DSM problem, the number of design variables is reduced to only 12 variables (6 DDVs and 6 CDVs). This step solves for the optimum mission's scenario without deep space maneuvers. Then, the second step is performed with 27 design variables

(5 DDVs and 22 CDVs). The ranges of the CDVs in the second step is reduced based on the information from the initial mission's scenario.

The ITO-HGGA tool optimizes also the number of revolutions in each leg. The time of flight of each leg and the time of applying each DSM are selected as design variables to allow for multi-revolution transfers. Lambert's problem is solved once in each leg. Based on the position vectors and the TOF, Lambert's solution may have both single and multi-revolution transfers. The selection criterion is as follows: the selected lambert's transfer should minimize the former maneuver cost. This maneuver could be a departure impulse, a DSM, or a swing-by maneuver. In the final leg, the selected lambert's transfer will minimize the former maneuver cost plus the arrival impulse cost.

The solution obtained by the genetic algorithm is not necessary an optimal solution, nor at a local minimum. Therefore, a constrained nonlinear optimization technique is used to improve the solution by finding the closest local minimum to that solution. The genetic algorithm solution is used as an initial guess in the local search algorithm.

9.2.2 Numerical Results

This section presents a number of case studies for interplanetary space missions trajectory optimization. Comparisons to other solutions in the literature is presented to validate the obtained results.

Earth to Mars (EM) Mission

A MGADSM trajectory is optimized for the Earth-Mars mission. The lower and upper bounds, for all design variables, are listed in **Table** 9.2. The maximum possible number of swing-by maneuvers is selected to be two. The maximum number of DSMs in each leg is selected to be two impulses. As can be seen from **Table** 9.2, a gravity assist maneuver could be performed with any planet in the solar system, from Mercury to Neptune. The direction of flight can be posigrade or retrograde. The time of flight of each leg, except the last one, is selected between 40 and 300 days.

The total number of design variables for this mission is 33 (7 DDVs and 26 CDVs). Wide ranges for the design variables are adopted, as listed in **Table** 9.2. A population of 500 individuals is used, for 300 generations. A local optimizer uses the fittest GA solution as an initial guess to find a local minimum. The resulting solution has a single swing-by maneuver at Venus, with a total cost of 10.754 km/s. The fittest trajectory has a single DSM in the first leg as shown in **Table** 9.3.

The same problem is solved again by splitting the optimization process into two steps to reduce the required computational time. First, a zero-DSM solution is sought to determine a mission scenario. The number of design variables in this

Table 9.2: Bounds of design variables for EM mission.

Design Variables	Lower Bound	Upper Bound
No. of swing-by maneuvers, N_s	0	2
Swing-by planets identification numbers, $P_1,..., P_i$	1 (Mercury)	8 (Neptune)
No. of DSMs in each mission's leg, $D_1,..., D_{i+1}$	0	2
Flight direction, F_d	Posigrade	Retrograde
Departure date, t_d	01-Jun-2004	01-Jul-2004
Arrival date, t_a	01-Apr-2005	01-Jul-2005
Time of flight (days)/leg, $T_1,..., T_i$	40	300
Swing-by pericenter altitude, $\bar{h}_1,..., \bar{h}_i$	0.1	10
Swing-by plane rotation angle (rad), $\eta_1,..., \eta_i$	0	2π
Epoch of DSM, $\varepsilon_1,..., \varepsilon_j$	0.1	0.9
Thrust impulse (km/sec), $\Delta\mathbf{v}_1,..., \Delta\mathbf{v}_k$	-5	5

Table 9.3: MGADSM solution trajectory for the EVM mission.

	MGADSM Scenario
Departure date, t_d	05-Jun-2004 00:28:38
Departure impulse (km/s)	4.53
DSM date	11-Sep-2004 18:35:44
DSM impulse (km/s)	0.1293 (t_d + 98.77 days)
Venus Swing-by date	19-Nov-2004 01:42:40
Post-swing-by impulse (km/s)	2.27e-12
Pericenter altitude (km)	7937.913
Arrival date	15-May-2005 15:07:38
Arrival impulse (km/s)	6.095
Time of flight (days)	167.07 , 177.6
Mission duration (days)	344.67
Mission Cost (km/s)	**10.754**

first step is only 8 variables. A genetic algorithm, with 200 populations and 100 generations, is used. The convergence of the optimization algorithm is shown in **Fig.** 9.4. The resulting zero-DSM scenario is a single swing-by maneuver at Venous with a cost of 10.788 km/s. Then, the local optimization tool is used to improve the solution to the nearest local minimum solution. The local optimizer reduces the cost to 10.783 km/s. A powered swing-by is implemented in this case and the required impulse of the swing-by is 0.002 km/s. **Figure** 9.5 shows the zero-DSM EVM trajectory.

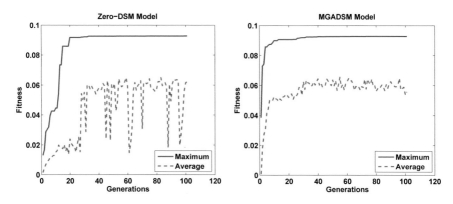

Figure 9.4: Convergence of the optimization algorithm for the EM mission using splitting technique.

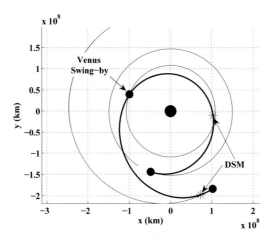

Figure 9.5: The solution trajectory for the EVM mission.

In the second step, the problem is solved using the full MGADSM formulation, assuming that the mission scenario is EVM (the scenario obtained from the first step). The ranges for the departure, swing-by, and arrival dates are varied within only 10 days around those values obtained from the first step. The number of design variables in this step is 12 (2 DDVs and 10 CDVs). A population of 300 members has been used for 100 generations of GA followed by a local minima optimizer. The result of the second optimization step is a single DSM of 180.1 m/s in the first leg at 80.81 days from mission start time, as shown in **Fig.** 9.5. The total transfer cost is slightly reduced to 10.728 km/s in the second optimization step. **Table** 9.4 shows the zero-DSM trajectory as well as the final trajectory. As shown in **Table** 9.4, the effect of the DSM is reducing the departure

Table 9.4: MGADSM solution trajectory for the EVM mission using splitting technique.

	Zero-DSM Model	MGADSM Model
	Initial Estimate	Final Scenario
Departure date,t_d	05-Jun-2004 01:52:21	02-Jun-2004 11:43:25
Departure impulse (km/s)	4.616	4.457
DSM date	-	22-Aug-2004 07:07:46
DSM impulse (km/s)	-	0.1801 (t_d + 80.81 days)
Venus Swing-by date	20-Nov-2004 15:10:59	19-Nov-2004 07:02:55
Post-swing-by impulse (km/s)	0.002	1.68e-13
Pericenter altitude (km)	7996.782	7869.048
Arrival date	14-May-2005 13:18:06	16-May-2005 03:29:08
Arrival impulse (km/s)	6.165	6.091
Time of flight (days)	168.56 , 174.92	169.81 , 177.85
Mission duration (days)	343.48	347.66
Mission Cost (km/s)	**10.783**	**10.728**

and arrival impulses, as well as reducing the impulse of the powered swing-by to almost zero. The fittest trajectory could be considered as a non-powered gravity assist trajectory with a single DSM.

Earth to Jupiter Mission (EJ)

A mission to planet Jupiter is considered. It is desired to find the MGADSM trajectory with minimum cost. Wide ranges of all design variables are implemented in optimization to explore more of the design space. The design variables' bounds are listed in **Table** 9.5. The total number of design variables is 33 variables (7 DDVs and 26 CDVs). The zero-DSM model is used to determine an initial scenario. The independent design variables in this step are 8 variables (4 DDVs and 4 CDVs). A population of 300 individuals and 100 generations are used in GA optimization. The resulting fittest scenario is a two swing-by trajectory with swing-bys around Venus then Earth (EVEJ), as shown in **Fig.** 9.7. The solution is a posigrade multi-revolution trajectory. The total cost of this zero-DSM trajectory is 10.298 km/s, as shown in **Table** 9.6. The scenario of this solution is then used in the second optimization step. The two swing-bys are fixed (EVEJ). The departure/arrival dates and the time of flight design variables obtained from the first step are allowed to vary within 10 days from their values obtained from the first step. A population of 500 individuals and 100 generations is used. **Figure** 9.6 shows the convergence of the optimization algorithm. **Table** 9.6 shows the zero DSM solution, as well as two solutions with DSMs. One solution has a single DSM in the second leg (VE), while the other solution has a single DSM in the first leg (EV), as shown in **Fig.** 9.8.

Table 9.5: Bounds of EJ mission's design variables.

Design Variables	Lower Bound	Upper Bound
No. of swing-by maneuvers, N_s	0	2
Swing-by planets identification numbers, $P_1,..., P_i$	1 (Mercury)	8 (Neptune)
No. of DSMs in each mission's leg, $D_1,..., D_{i+1}$	0	2
Flight direction, F_d	Posigrade	Retrograde
Departure date, t_d	01-Sep-2016	30-Sep-2016
Arrival date, t_a	01-Sep-2021	31-Dec-2021
Time of flight (days)/leg, $T_1,..., T_i$	80	800
Swing-by pericenter altitude, $\bar{h}_1,..., \bar{h}_i$	0.1	10
Swing-by plane rotation angle (rad), $\eta_1,..., \eta_i$	0	2π
Epoch of DSM, $\varepsilon_1,..., \varepsilon_j$	0.1	0.9
Thrust impulse (km/sec), $\Delta v_1,..., \Delta v_k$	-5	5

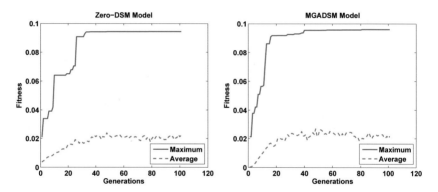

Figure 9.6: Convergence of the optimization algorithm of Jupiter mission.

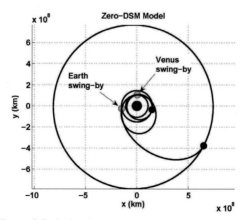

Figure 9.7: Optimal zero-DSM trajectory of EJ mission.

Table 9.6: Optimal MGADSM trajectory of EVEJ mission.

	Zero-DSM Model	MGADSM Model	
	Initial Estimate	Scenario (1)	Scenario (2)
Departure date	09-Sep-2016 11:38:03	06-Sep-2016 13:36:17	07-Sep-2016 01:55:17
Departure impulse (km/s)	3.653	3.542	3.439
DSM date	-	-	21-Feb-2017 08:29:43
DSM impulse (km/s)	-	-	0.109
Venus swing-by date	05-Sep-2017 05:57:07	05-Sep-2017 14:57:28	07-Sep-2017 07:43:57
Post-swing-by impulse(km/s)	0.0004	-	2.38e-014
Pericenter altitude (km)	1402.2	1307.28	613.545
DSM date	-	14-May-2018 09:31:08	-
DSM impulse (km/s)	-	0.0002	-
Earth swing-by date	30-Mar-2019 02:25:00	30-Mar-2019 03:14:06	29-Mar-2019 02:19:05
Post-swing-by impulse(km/s)	0.443	0.441	0.444
Pericenter altitude (km)	637.8	637.8	637.8
Arrival date	18-Sep-2021 21:15:27	24-Sep-2021 23:59:59	17-Sep-2021 07:43:51
Arrival impulse (km/s)	6.202	6.195	6.19
Time of flight (days)	360.76, 570.85, 903.79	364.05, 570.51, 909.87	365.24, 567.77, 903.23
Mission duration (days)	1835.4	1844.43	1836.24
Mission Cost (km/s)	**10.298**	**10.178**	**10.182**

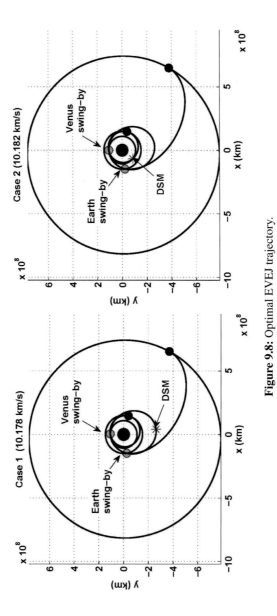

Figure 9.8: Optimal EVEJ trajectory.

Earth to Saturn Mission (Cassini 2)

One of the most complicated multi gravity assist trajectories is the Cassini 2 trajectory. In 1997, Cassini-Huygens mission was launched to study the planet Saturn and its moons. The ITO-HGGA tool is used to search for a minimum cost trajectory for an Earth-Saturn trip. For the sake of making comparisons with the literature, a narrow range for departure date is selected, around the known published date for the Cassini 2 mission. The ranges for the other design variables are wide enough to investigate all possible solutions. **Table** 9.7 presents the upper and lower bounds for the design variable. A zero-DSM model is initially solved to calculate an initial mission scenario. The independent design variables in this step are 12 variables (6 DDVs and 6 CDVs). A population of 500 members and 1000 generations are used. The GA convergence is shown in **Fig.** 9.9.

Table 9.7: Bounds of Cassini 2 mission's design variables.

Design Variables	Lower Bound	Upper Bound
No. of swing-by maneuvers, N_s	1	4
Swing-by planets identification numbers, $P_1,..., P_i$	1 (Mercury)	8 (Neptune)
No. of DSMs in each mission's leg, $D_1,..., D_{i+1}$	0	1
Flight direction, F_d	Posigrade	Retrograde
Departure date, t_d	01-Nov-1997	31-Nov-1997
Arrival date, t_a	01-Jan-2007	30-Jun-2007
Time of flight (days)/leg, $T_1,..., T_i$	40	1000
Swing-by pericenter altitude, $\bar{h}_1,..., \bar{h}_i$	0.1	10
wing-by plane rotation angle (rad), $\eta_1,..., \eta_i$	0	2π
Epoch of DSM, $\varepsilon_1,..., \varepsilon_j$	0.1	0.9
Thrust impulse (km/sec), $\Delta v_1,..., \Delta v_k$	-5	5

The resulting scenario is a four swing-by maneuver with the same planet sequence as of the actual cassini 2 mission scenario (EVVEJS). The trajectory is a posigrade multi revolution transfer with 10.836 km/s total transfer cost, as shown in **Table** 9.8. Then, the initial zero-DSM scenario is used in the MGADSM model to obtain the final trajectory. The planet sequence is fixed at (EVVEJS), with narrow ranges for the departure, arrival, and gravity assist maneuvers dates. The population size is selected to be 500 members and the number of generations is 1000. Then, a local optimizer is used. The number of design variables is 27 in this model (5 DDVs and 22 CDVs). The GA convergence is shown in **Fig.** 9.9. The final solution is presented in **Table** 9.8. The trajectory has two deep space maneuvers, one in the first leg (EV) and the other is in the second leg (VV). The

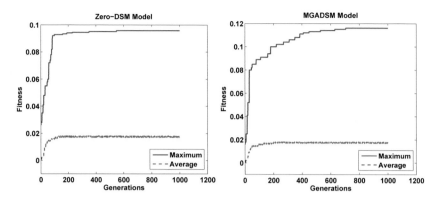

Figure 9.9: Genetic algorithm performance of Cassini 2 mission.

total transfer cost is reduced to 8.385 km/s after applying the deep space maneuvers. The zero-DSM initial trajectory and the MGADSM final trajectory are shown in **Fig.** 9.10.

9.2.3 *Discussion*

The cost that is being optimized in this work is a function of discreet and continuous design variables. The discrete variables determine the mission scenario (swing-by planets), the number of DSMs in each leg, and the direction of flight. The continuous variables determine times of all impulses and their amounts. The ITO-HGGA tool implemented in this chapter finds the values for all the design variables in the most obtainable fit solution trajectory. Then the mission scenario, the direction of flight, and the number of DSMs are fixed, while the continuous variables are tuned by a local optimizer. The local optimizer implements a constrained nonlinear optimization technique.

The selection for the GA parameters depends on the specific problem. The crossover probability is varied from 0.9 to 0.98, and the mutation probability is selected between 0.01 and 0.08. Proportional ranking is implemented in most cases. In general, linear ranking is implemented in association with higher crossover probability values, in solving a zero-DSM problem, in order to increase the diversity in the population. The linear ranking is implemented to avoid local minima traps.

The Earth-Mars mission trajectory optimization has been addressed in the literature [79]. The solution presented in Ref. [79] was obtained using the extended primer vector theory, and has a single swing-by maneuver and a single DSM. In implementing the primer vector method, the departure and arrival dates were assumed fixed (the mission duration is 340 days). The Venus swing-by time was also constrained to occur at 165 days from departure. The resulting solution has

Table 9.8: Optimal MGADSM trajectory of Cassini 2 mission.

	Zero-DSM Model Initial Estimate	MGADSM Model Final Scenario
Departure date	14-Nov-1997 18:34:18	13-Nov-1997 11:12:37
Departure impulse (km/s)	3.783	3.293
DSM date	-	25-Mar-1998 22:27:03
DSM impulse (km/s)	-	0.449
Venus Swing-by date	13-May-1998 10:54:34	30-Apr-1998 04:17:03
Post-swing-by impulse (km/s)	2.013	-
Pericenter altitude (km)	21094.164	2590.174
DSM date	-	11-Dec-1998 14:55:31
DSM impulse (km/s)	-	0.396
Venus Swing-by date	26-Jun-1999 17:44:57	27-Jun-1999 11:24:49
Post-swing-by impulse (km/s)	0.789	2.21e-06
Pericenter altitude (km)	3622.073	245.216
Earth Swing-by date	19-Aug-1999 11:26:01	19-Aug-1999 16:18:04
Post-swing-by impulse (km/s)	2.22e-11	6.04e-08
Pericenter altitude (km)	1865.136	1975.905
Jupiter Swing-by date	29-Mar-2001 20:33:42	31-Mar-2001 09:30:36
Post-swing-by impulse (km/s)	1.07e-05	1.68e-07
Pericenter altitude (km)	4973853.7	4918886.8
Arrival date	22-Mar-2007 18:12:55	23-Mar-2007 21:31:17
Arrival impulse (km/s)	4.251	4.247
Time of flight (days)	179.68, 409.29, 53.74, 588.38, 2183.9	168.67, 423.3, 53.2, 589.72, 2182.54
Mission duration (days)	3414.99	3417.43
Mission Cost (km/s)	10.836	8.385

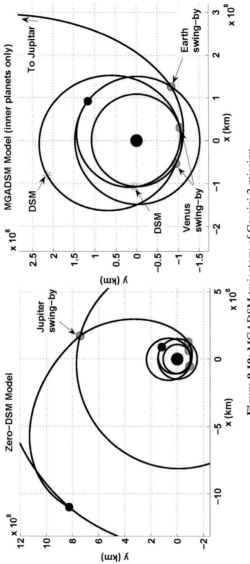

Figure 9.10: MGADSM trajectory of Cassini 2 mission.

a DSM of 68.7 m/s in the first leg at 96.08 days from mission start date. The total cost of the mission is 10.786 km/s [79]. The solution obtained using the ITO-HGGA tool in this chapter has also one swing by Venus and one DSM of 180.1 m/s in the first leg at 80.81 days from launch date. The total cost of the mission is 10.728 km/s, as shown in **Table** 9.4. The reduction in the total cost obtained using the ITO-HGGA tool, as compared to that of reference [79], is accompanied by changes in the maneuver's launch, swing-by, and arrival dates, without significantly changing the total mission duration. This solution is obtained by allowing the maneuver's launch, swing-by, and arrival dates to be freely chosen during the zero-DSM optimization step.

Reference [79] presents also a minimum cost solution trajectory for the Earth-Jupiter mission, assuming the fixed planet sequence EVEJ. The departure, arrival, and swing-by dates were also assumed fixed, with a launch in 2016 and a mission duration of 1862 days. The primer vector theorem solution has four DSMs. Two DSMs are applied in the first two legs. The total transfer cost for this solution is 10.267 km/s. The ITO-HGGA tool presented in this chapter is able to find automatically the known swing-by sequence for the Jupiter mission in 2016. As listed in **Table** 9.6, the initial scenario (zero-DSM trajectory) is obtained, starting from wide ranges for the design variables. Then, the MGADSM solutions are obtained based on that initial scenario. Each of the two MGADSM solutions , presented in **Table** 9.6, have less total cost as compared with the solution presented in reference [79]. The number of DSMs in each of the two solutions is only one DSM. The solution presented in reference [79] requires four DSMs.

The Cassini 2 mission trajectory design problem has been addressed in several studies [79, 90, 115], where it is always assumed that a fixed swing-by sequence (EVVEJS) is known. In reference [115], the IMAGO obtained a solution that has a single DSM, and a total cost of 9.06 km/s. Reference [79] implements the space pruning technique and finds the values and locations of the DSMs to minimize the total transfer cost. The departure, arrival, and swing-by dates were fixed. The minimum cost obtained in [79] is 8.877 km/s. Olympio and Izzo [90] recently developed an algorithm to find the optimal DSM structure in the a given trajectory scenario. For the Cassini 2 planets' sequence described in the GTOP database [41], reference [90] found a trajectory with a total cost of 8.387 km/s. The solution obtained using the ITO-HGGA tool in this chapter finds a solution with a total cost of 8.385 km/s (see **Table** 9.8), which is very close to the reported result in the GTOP database (8.383 km/s) [41], and is also very close to the solution presented in reference [90]. The ITO-HGGA tool in this case, however, finds the planets sequence as well as the DSM structure.

9.3 Trajectory Optimization using HGGA with Binary Tags

This section presents the implementation of the HGGA with binary tags to the space trajectory optimization problem. A space mission from Earth to Jupiter is first considered as a case study. The results presented in this section were first published in reference [5].

9.3.1 Earth–Jupiter Mission using HGGA

The Earth-Jupiter mission is optimized using the HGGA and the results are compared to other known solutions in the literature. The design variables and their given upper and lower bounds are listed in **Table** 9.9. In this test, it is assumed that the maximum number of fly-bys is 2. Hence the chromosome has 2 genes for the fly-by planets; each gene carries the planet identification number (the planet identification number ranges from 1 to 8 for all the planets in the solar system as shown in **Table** 9.9). Each of these two genes has a tag. If both tags have a value of 1, then the two genes are hidden and that solution has no fly-bys. If one gens is hidden and one is active, then the solution has one fly-by and the fly-by planet identification number is the value of that gene. The same concept is applied regarding the DSMs. In this test case it is assumed that the maximum number of DSMs in each leg is 2. Since the maximum number of fly-bys is 2 then the maximum number of legs is 3, and hence the maximum number of DSMs is 6. For each DSM, we need to compute the optimal time (TD) at which this DSM occurs. A gene and a tag are added for each DSM time TD, and hence there are 6 genes and 6 tags for TD_i $(i = 1 \cdots 6)$ in this mission. Note that if a fly-by is hidden, then its leg disappears and the DSMs in that leg automatically become hidden. Note also that even if a fly-by exists, a DSM in its leg can be hidden depending on the value of its own tag. Each DSM is an impulse represented by a vector of three components (it has the units of velocity). So, the chromosome will have genes for $6 \times 3 = 18$ scalar components of the DSMs. Note that these 18 genes are grouped in groups of three genes since each three are the components of one DSM vector; hence if one DSM is hidden then its 3 genes get hidden together. **Table** 9.9 shows the ranges for 6 DSMs, each has 3 components. The Time Of Flight (TOF) in each leg is also a variable; there is a gene for each TOF in the mission. Hence in this Jupiter mission, we have 3 genes for the TOFs. Note that there are no tags associated with the TOF genes since the state of each gene (hidden or active) is determined based on the fly-by tags. If a fly-by exists then there is an active gene for a TOF associated with it. There is at least one TOF in a mission; this case of only one TOF corresponds to the case of no fly-bys. To complete a fly-by maneuver, we need to calculate the optimal values for the normalized altitude h of the spacecraft above the planet as well as the plane angle η of the maneuver. Hence, two genes for the altitudes and two genes for the

Table 9.9: Lower and upper bounds of Earth-Jupiter problem.

Design Variable	Lower Bound	Upper Bound
planet of Fly-by #1	1 (*Mercury*)	8 (*Neptune*)
planet of Fly-by #2	1	8
Epoch of DSM (TD$_i$), $i = 1 \cdots 6$	0.1	0.9
DSM$_i$ vector (km/s), $i = 1 \cdots 6$	$[-5, -5, -5]$	$[5, 5, 5]$
TOF$_i$ (days), $i = 1 \cdots 3$	80	800
Flyby altitudes h_i, $i = 1 \cdots 3$	0.1	10
Flyby angle (rad) η_i, $i = 1 \cdots 3$	0	2π
Departure Impulse (km/s)	$[-5, -5, -5]$	$[5, 5, 5]$
Flight Direction	Posigrade	Retrograde
Departure Date	01 Sep.2016	30 Sep.2016
Arrival Date	01 Sep.2021	31 Dec.2021

angles are added. Similar to the TOF variables, no tags are needed for the h and η genes. There are also 6 genes for the departure impulse, flight direction, the arrival date and the departure date.

Due to the high computational cost of evaluating the fitness of a solution and hence the high computational cost of searching for the optimal solution, the problem is usually solved in two steps [49]. The first step is to assume no DSMs in the trajectory, and this step is called zero-DSM. In the second step, the fly-by planets sequence is obtained from the first step, and held fixed, to search for the optimal values of the remaining variables; this step is called the Multi Gravity Assist with Deep Space Maneuvers (MGADSM). Each of the two steps has an element of the architecture to be optimized: the first step optimizes the sequence of fly-bys and the second step optimizes the number of DSMs in each leg. This two-step approach is detailed in reference [49], and has shown to be computationally efficient. The number of generations and population size are selected to be 100 and 300, respectively.

9.3.2 Earth–Jupiter Mission: Numerical Results and Comparisons

All mechanisms evolving the tags over subsequent generations were tested on the Earth-Jupiter problem. Mechanism A (with a mutation probability of 5%) generated the best solution and so its solution is presented here in detail. The solution of the first phase shows that the trajectory consists of 2 fly-bys around planets Venus and then Earth; the mission sequence is then Earth-Venus-Earth-Jupiter (EVEJ). The second phase results in adding one DSM, and the total mission cost is 10.1266 km/sec (the fuel consumption can be measured in velocity units). The resulting mission is detailed in **Table** 9.10.

Table 9.10: HGGA solution of Earth-Jupiter problem using Mechanism A.

Mission parameter	Zero-DSM model (first step)	MGADSM model (second step)
Departure Date	$11 - Sep - 2016, 04 : 50 : 27$	$31 - Aug - 2016, 02 : 15 : 15$
Departure Impulse (km/s)	3.6283	3.4414
DSM_1 date	—	$26 - Oct - 2016, 21 : 29 : 29$
DSM_1 impulse (km/s)	—	0.036516
Venus flyby date	$07 - Sep - 2017, 01 : 58 : 51$	$05 - Sep - 2017, 09 : 52 : 46$
Post-flyby impulse (km/s)	0.031108	0.001704
Pericenter altitude (km)	1333.7876	1255.2411
Earth flyby date	$03 - Apr - 2019, 09 : 55 : 30$	$29 - Mar - 2019, 09 : 56 : 34$
Post-flyby impulse (km/s)	0.4685	0.4522
Pericenter altitude (km)	637.7999	637.8000
Arrival date	$25 - Dec - 2021, 23 : 05 : 00$	$25 - Dec - 2021, 18 : 23 : 09$
Arrival impulse (km/s)	6.2813	6.1948
TOF (days)	360.88083, 934.21184, 636.66743	370.3177, 570.0026, 902.3518
Mission cost (km/s)	10·4092	10·1266

Also both Logic C and Logic A were tested on this problem; logic A demonstrated superiority compared to logic C in this problem and hence only the results of logic A are here presented. The total cost for the mission is 10.1181 km/sec. The detailed results of both steps are presented in **Table** 9.11.

Table 9.11: HGGA solution of Earth-Jupiter problem using Logic A.

Mission parameter	Zero-DSM model (first step)	MGADSM model (second step)
Departure Date	$01 - Sep - 2016, 22:58:19$	$29 - Aug - 2016, 16:04:38$
Departure Impulse (km/s)	3.4811	3.1398
DSM date	–	$07 - Oct - 2016, 05:57:10$
DSM impulse (km/s)	–	0.34746
Venus flyby date	$05 - Sep - 2017, 10:42:20$	$06 - Sep - 2017, 19:14:50$
Post-flyby impulse (km/s)	0.006200	$1.9788e - 05$
Pericenter altitude (km)	1330.9042	972.1739
Earth flyby date	$29 - Mar - 2019, 23:48:29$	$29 - Mar - 2019, 14:31:32$
Post-flyby impulse (km/s)	0.4398	0.4396
Pericenter altitude (km)	637.8000	637.8000
Arrival date	$19 - Sep - 2021, 03:07:38$	$23 - Sep - 2021, 15:01:36$
Arrival impulse (km/s)	6.1999	6.1891
TOF (days)	368.4889,570.5459,904.1383	373.1321,568.8033,909.0209
Mission cost (km/s)	10.1299	10.1181

Comparing the results of Logic A with Mechanism A, we can see that the cost of the mission using Logic A is slightly better than that obtained using Mechanism A. The mission architecture is the same from both methods, while the values of the other variables are slightly different. The success rate of an algorithm is a measure for how many times the algorithm finds the best found solution in a repeated experiment (see **Chapter** 6). This experiment was repeated 200 times using Logic A and the success rate is 75.5%, as shown in **Fig.** 9.11.

Previous solutions in the literature for this problem can be divided into two categories. The first category of methods do not search for the optimal architecture; rather the trajectory is optimized for a given architecture. Reference [91], for instance, presents a minimum cost solution trajectory for this Earth-Jupiter mission, assuming a fixed planet sequence of EVEJ. The departure, arrival, and fly-bys dates were also assumed fixed, with a launch in 2016 and a mission duration of 1862 days. The primer vector theorem solution has four DSMs. Two DSMs are applied in the first two legs. The total cost for this solution is 10.267 km/s, which is about slightly higher than the obtained cost in this section. The method presented in this section, however, has the advantage of the autonomous search for the optimal architecture of the solution. The obtained solution here has the same planet sequence of EVEJ but a different DSM architecture compared to [91]. Reference [49] and **Section** 9.2 present the solution to this problem using the simple HGGA (without the tags concept). The solution in **Section** 9.2 also finds the planet sequence of EVEJ, and has a total mission cost of 10.182 km/sec, which is slightly higher than the cost obtained in this section. This problem was also solved using Mechanisms E and F (presented in **Section** 6.5) and the results were presented in [4]. The total cost obtained using Mechanism E is 10.1438

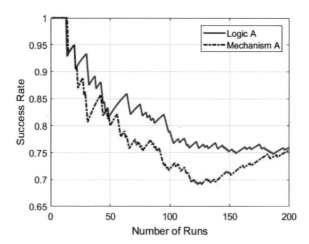

Figure 9.11: Success rate versus number of runs for the Earth-Jupiter trajectory optimization problem using Mechanism A and Logic A.

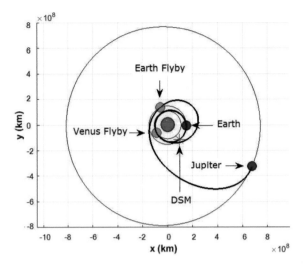

Figure 9.12: Mechanism A: EVEJ Trajectory for MGADSM Model.

km/s and using Mechanism F is 10.9822 km/s, which are higher than the cost obtained in this section. The mission trajectory obtained using Mechanism A is shown in **Fig.** 9.12.

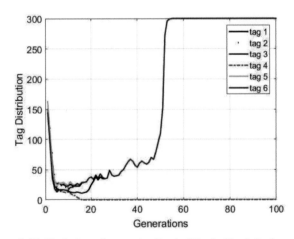

Figure 9.13: Evolution of tags using Logic C in the Earth-Jupiter problem.

As a demonstration for how the tags evolve over subsequent generations, consider this Earth-Jupiter problem solved using Logic C. The population size is 300 and the number of generations is 100. Six tags are examined. **Figure** 9.13 shows the number of times each tag has a value of '1' in each generation. For example, tag 6 takes a value of '1' in all the population members in generations 55 and above. In the 30th generation, for instance, tag 6 takes a value of '1' in only 40 chromosomes and takes a value of '0' in the other 260 chromosomes. The other 5 tags converge to a value of '0' in the last population in all the chromosomes.

Chapter 10

Control and Shape Optimization of Wave Energy Converters

In some applications, the simultaneous optimization of both the system design and the control is crucial for optimizing the system operations. Examples of such applications include the design and control of offshore wind turbines, and ocean wave energy converters. The latter application is discussed in some detail in this chapter. The results presented in this chapter were first published in reference [6].

One of the challenges in wave energy harvesting is the motion control. There have been significant developments for different control methods for WECs. Most studies on the control of one-degree-of-freedom heaving wave energy converters adopt a linear dynamic model—the Cummins' equation—which can be written as:

$$(m+\tilde{a})\ddot{z} = F_{ex} + u - B_v\dot{z} - kz - \int_0^\infty h_r(\tau)\dot{z}(t-\tau)d\tau \qquad (10.1)$$

where z is the heave displacement, m is the buoy mass, k is the hydrostatic stiffness due to buoyancy, $\tilde{a}$ is the added mass, F_{ex} is the excitation force, u is the control force, B_v is a viscous damping coefficient, and h_r is the radiation impulse response function (radiation kernel). The radiation term is called radiation force, F_r, and the buoyancy stiffness term is called the hydrostatic force.

There are multiple sources of possible nonlinearities in the WEC dynamic model though.For example, if the buoy shape is not a vertical cylinder near the water surface, then the hydrostatic force will be nonlinear [60]. The hydrodynamic forces can also be nonlinear in the case of large motion. Control strategies that aim at maximizing the harvested energy usually increase the motion amplitude and, hence, increase the impact of these nonlinearities.

In the case of using nonlinear control, it is possible that the motion of the buoy grows large enough to make the buoy almost fully submerged or almost fully in the air. In such cases, the linear hydrodynamic model becomes invalid, and modeling of nonlinear hydrodynamics becomes inevitable. In this section, we examine the impact of having nonlinear terms in the equation of motion whether they appear due to nonlinear hydrodynamics, nonlinear hydrostatics, nonlinear damping, and/or nonlinear control forces. Toward that goal, the control force is here assumed in the form of a summation of two quantities:

$$u = u_l + \tilde{u}_c \tag{10.2}$$

where u_l is the linear part of the control and $\tilde{u}_c$ is the nonlinear control part. The harvested power can be expressed as:

$$P(t) = -(\tilde{u}_c(t) + u_l) \times \dot{z}(t) = -u \times \dot{z}(t) \tag{10.3}$$

The nonlinear control part is assumed in the form:

$$\tilde{u}_c = \sum_{i=1}^{Nc} \alpha_{c_i} z^i + \sum_{j=1}^{Mc} \beta_{c_j} \dot{z}^j \tag{10.4}$$

where α_{c_i} and β_{c_j} are constant coefficients and N_c and M_c are the number of nonlinear terms that determine the order of control forces.

The hydrodynamic and hydrostatic (hydro) forces along with all other optimizable nonlinear forces are referred to as the system nonlinearities. The system nonlinearities and the control (also nonlinear) are here optimized simultaneously. For the sake of control design, it is convenient to express the optimizable system nonlinearities as a series function as follows:

$$\tilde{f}_s = \sum_{i=1}^{Ns} \alpha_{s_i} z^i + \sum_{j=1}^{Ms} \beta_{s_j} |\dot{z}^j| sign(\dot{z}) \tag{10.5}$$

where $\tilde{f}_s$ is the nonlinear force, α_{s_i} and β_{s_i} are constant coefficients, $\forall i$; N_s and M_s are the number of nonlinear terms that determine the order of the nonlinear forces. Equation (10.5) is written intuitively; consider for example the Proportional-Derivative (PD) controls which are widely used in linear systems. In a PD control, the proportional part is constructed as linear term in the state, and the derivative term is constructed as a linear term in the state derivative. The proportional term is a stiffness term since it has spring-like effect, which means this part of the force does not add/remove energy on average. The derivative term, however, is a damper-like term, and it continuously adds/removes power. One might think of nonlinear stiffness or damping terms, as discussed in details in several references such as [85]. The first term in $\tilde{f}_s$ represents a nonlinear stiffness force, and the second term contains a nonlinear damping force. Note that all β_{s_j} are always negative coefficients, and hence the second term is always a damping term (energy flow is always from the water to the device). Optimizing the system nonlinearities means in this case finding the optimal coefficients α_{s_i}, $\forall i = 1 \cdots Ns$, and β_{s_i}, $\forall i = 1 \cdots Ms$. Similarly, optimizing the control means finding the optimal α_{c_i}, $\forall i = 1 \cdots Nc$ and β_{c_i}, $\forall i = 1 \cdots Mc$. Once $\tilde{u}_c$ and $\tilde{f}_s$ are optimized, the WEC system (e.g., the buoy shape) is designed so that the WEC nonlinear force matches the optimized nonlinear force $\tilde{f}_s$. This last step of designing a WEC system to generate a prescribed nonlinear force is not addressed in this chapter; the focus of this chapter is on the optimization of $\tilde{f}_s$ and $\tilde{u}_c$.

The equation of motion of the system then is:

$$(m + \tilde{a}_\infty)\ddot{z} + B_v\dot{z} + kz = f_{ex} + f_r + u_l + \tilde{u}_c + \tilde{f}_s \tag{10.6}$$

The equation of motion, Eq. (10.6), is derived assuming that the buoy does not leave the water nor gets fully submerged in the water. In the case of nonlinear WECs presented in this chapter, the motion of the buoy may grow large and these two cases should not be excluded. Hence the model in Eq. (10.6) is modified as follows. Consider the coordinates defined in **Fig.** 10.1, a range $|z| < z_s$ is defined in which the model in Eq. (10.6) is considered valid. The limit z_s is selected based on the buoy dimensions and the wave height. When $|z| > z_s$, there are two possible cases. The first case is when $(z > 0)$, that is the buoy is (or very close to being) fully submerged under water. The second case is when $(z < 0)$, that is the buoy is (or very close to being) totally out of the water. In these two cases, the dynamic model in Eq. (10.6) is not valid, and an approximate dynamic model is defined as follows:

Case 1: $(z > 0)$ The linear stiffness term becomes a constant $kh/2$. The nonlinear stiffness force will also be saturated, that is $\tilde{f}_s = \sum_{i=2}^{Ns} \alpha_{s_i}(h/2)^i + \sum_{j=1}^{Ms} \beta_{s_j}|\dot{z}^j|sign(\dot{z})$. The excitation force is assumed to remain the same as in Eq. (2.26).

Case 2: $(z < 0)$ The buoy is out of the water so there is no buoyancy force on it, meaning that $kz = -mg$, where g is the gravitational acceleration. There is no excitation force acting on the WEC; and there is no linear damping term in Eq. (10.6). Also the nonlinear hydro force vanishes, that is $\tilde{f}_s = 0$. The equation of motion reduces to $m\ddot{z} = mg + u_l + \tilde{u}_c$.

The harvested power $P(t)$ is expressed as:

$$P(t) = -(\tilde{u}_c(t) + u_l) \times \dot{z}(t) = -(u) \times \dot{z}(t) \tag{10.7}$$

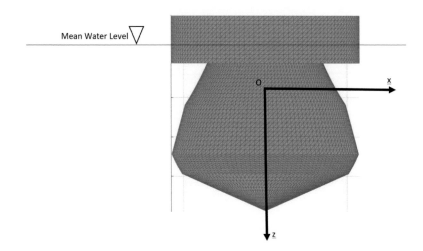

Figure 10.1: Buoy coordinate system.

In the analysis conducted in this study, u_l is assumed a damping force as follows:

$$u_l = B\dot{z} \tag{10.8}$$

where B is the linear control force damping coefficient; B is negative. The shape and control coefficients will be optimized so as to maximize the harvested energy. The optimization problem can be formulated as follows:

$$\text{Maximize: } E = \int_0^T P(t)dt \tag{10.9}$$

$$\text{Subject to: Eq. (10.6).}$$

where E is the energy harvested over a period T, which can be expressed as:

$$E = \int_0^T P(t)dt = -\int_0^T u\dot{z}dt = -\int_0^T \dot{z}\left(B\dot{z} + \sum_{i=1}^{Nc}\alpha_{c_i}z^i + \sum_{j=2}^{Mc}\beta_{c_j}\dot{z}^j\right)dt \tag{10.10}$$

The design variables in this optimization problem are N_s, M_s, N_c, M_c, α_{s_i}, β_{s_j}, $\forall$ $i = 1, ..., N_s$ and $j = 1, ..., M_s$, α_{c_k}, and β_{c_l}, $\forall$ $k = 1, ..., N_c$ and $l = 1, ..., M_c$. The overall number of variables is variable since N_s, M_s, N_c, and M_c are variables, and hence this is a VSDS optimization problem. This problem can be solved using HGGA, SCGA, or DSMPGA. A numerical case study is presented below.

10.1 A Conical Buoy in Regular Wave

This chapter focuses on the optimization of the nonlinear control and system forces without addressing how the optimized system forces can be realized. This example, however, is presented to demonstrate how a buoy shape can be optimized along with the control. Consider a buoy of conical shape. Reference [58] presents an efficient way of computing the static and dynamic nonlinear Froude-Krylov forces, and shows that these nonlinear Froude-Krylov forces reduces the amount of harvested energy compared to the linear system. Here it is shown that when a nonlinear control is optimized along with the shape, a significant increase in the harvested energy can be achieved. Assume that the system nonlinear forces $\tilde{f}_s$ are only those due to the nonlinear Froude-Krylov forces. Two variables are used in optimizing the cone dimensions: the cone angle (θ) and the buoy radius at water level (r_w). For simple presentation here we assume that $\tilde{f}_s(\theta, r_w)$ is a function of only θ and r_w. The HGGA presented in **Section** 6.3 is used to optimize the control force $\tilde{u}_c$ and the buoy shape parameters θ and r_w. In this case the number of control variables is variables while the number of buoy shape variables is fixed (2 variables). **Table** 10.1 lists the ranges selected for the design variables where the number of control variables can be up to 12. A regular wave is assumed that has an amplitude of 0.3 m and a period of 9 s. No viscous damping is assumed. During the optimization process, the HGGA tries many solutions of different selections of the design variables. For each set of the design variable, the boundary element solver NEMOH is used to compute the hydrodynamic coefficients needed to compute the linear diffraction and radiation forces. Then a motion numerical simulation is conducted using the model described in Eq. (10.6). At each time step in the numerical simulation, the nonlinear static Froude-Krylov force $f_{FK_{st}}(t)$ and the nonlinear dynamic Froude-Krylov force $f_{FK_{dyn}}(t)$ are computed for this conical buoy as function of the system states and the design variables as follows [58]:

$$f_{FK_{st}}(t) = f_g - 2\pi\rho g m \left[m \frac{\sigma^3}{3} + (q - mz) \frac{\sigma^2}{2} \right]_{\sigma_1}^{\sigma_2} \quad (10.11)$$

Table 10.1: Conical Buoy: bounds on design variables.

Design Variable	Value
$\alpha_{c_{min}}$	$-1000 \times [1,1,1,1,1,1]$
$\alpha_{c_{max}}$	$1000 \times [1,1,1,1,1,1]$
$\beta_{c_{min}}$	$-1000 \times [1,1,1,1,1,1]$
$\beta_{c_{max}}$	$1000 \times [1,1,1,1,1,1]$
$[\theta, r_w]_{min}$	$[10°, 0.5\text{m}]$
$[\theta, r_w]_{max}$	$[30°, 1\text{m}]$

$$f_{FK_{dyn}}(t) = \frac{2\pi}{\chi}\rho gam^2 cos(\omega t)\left[\left(\frac{q}{m}-\frac{1}{\chi}-z+\sigma\right)e^{\chi\sigma}\right]_{\sigma_1}^{\sigma_2} \qquad (10.12)$$

where, $\sigma_1 = z - h_0$, σ_2 is the free surface elevation, h_0 is the buoy draft under water in rest position (still water), $\chi = \dfrac{\omega^2}{g}$ is the wavelength in deep water, ω is the wave frequency, g is the gravitational acceleration, and $f_g = -mg$. The nonlinear control force is also evaluated at each time step as function of the current states and the design variables. The HGGA process evolves until a convergence is achieved. The results of optimization yields a buoy shape of radius $r_w = 0.578$ m and an angle of $\theta = 16.5°$ (that means the draft is 1.95 m), and a nonlinear control of the form:

$$\tilde{u}_c = 669.5z + 456.4z^2 + 783.7z^3 - 999.9\dot{z} - 488.7\dot{z}^2 \qquad (10.13)$$

The results of the nonlinear controller are compared to two linear controllers; the linear damping controller and the Proportional Derivative Complex Conjugate Controller (PDC3) [13, 105]. The PDC3 control is a time-domain control that approximates the complex conjugate control; the PDC3 has both stiffness and damping terms and hence it can provide has reactive power, unlike the linear damping control. The PDC3 control is detailed in references [13, 105]; it is designed based on a linear model for the WEC. Yet it is here tested in a nonlinear environment, using a propagator that has nonlinear Froude-Krylov force, to highlight the significance of having a nonlinear control force. For the linear damping controller, the damping coefficient is selected to be equal to β_{c1} in the nonlinear controller. The PDC3 has stiffness and damping terms as follows:

$$u_{PDC3} = -k_{up}z - k_{ud}\dot{z} \qquad (10.14)$$

The cone buoy, with the shape that is obtained from optimization, is simulated using a nonlinear model; the harvested energy of the nonlinear and linear controllers are shown in **Fig.** 10.2. The results show the harvested energy increases by a factor of 3.5 in the case of the nonlinear controller compared to the linear controllers. This result emphasizes the importance of designing a nonlinear control when the there are nonlinear hydrodynamic forces. In linear WECs (linear hydrodynamic forces and cylindrical buoy shape), the PDC3 generates higher energy compared to the linear damping control. In fact, reference [14] presents a linear WEC case study in which the complex conjugate control generates about 3.5 times the energy harvested using a linear damping control, which is about the same ration this nonlinear control achieves in the case of a nonlinear WEC. In the steady state, the nonlinear control force has a maximum value of about 2500 N while the linear one has a maximum of about 250 N for the damping controller and 1000 N for the PDC3 controller. The displacement of the buoy in the nonlinear case has a maximum of 1.17 m while in the linear case it is 0.65

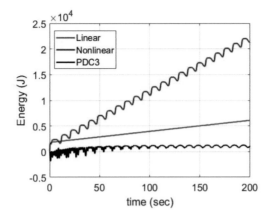

Figure 10.2: Energy harvested using a nonlinear controller is higher than that of the linear controllers for the same conical buoy shape when optimizing the control and shape in the nonlinear case.

m for the damping controller and 2.25 m for the PDC3 controller. The maximum buoy velocity in the nonlinear case is 1.6 m/s while in the linear case it is 0.25 m/s for the linear damping controller and 4.0 m/s for the PDC3 controller.

10.2 General Shape Buoys in Regular Waves

This section presents a simple illustrative numerical example to highlight the advantages that can be attained when optimizing both the system and control nonlinear forces. Consider the case of a system described by Eq. (10.6), that has optimizable nonlinear forces. Assume a buoy mass of 1.76×10^5 kg. The linear stiffness force coefficient is 4.5449×10^5 kg/s^2, the linear damping coefficient is 170 kg/s, the linear control force damping coefficient is 4×10^4 kg/s, the reference amplitude of excitation force is 263980 N, its frequency is 0.628 rad/s, and its phase is 0.1 rad. The buoy height is assumed 7.6 m, and the limit z_s is 3.4 m. For the nonlinear part, the number of nonlinear terms in the system nonlinear force is selected as: $N_s = 4$ and $M_s = 2$. The nonlinear hydro force coefficients are:

$$\alpha_s = [0, 11556.5594, 8565.3620, -1296.1921] \tag{10.15}$$
$$\beta_s = [-13925.7309, -550011.5422] \tag{10.16}$$

The number of control terms are selected as: $N_c = 4$ and $M_c = 3$. The nonlinear control coefficients are:

$$\alpha_c = [0, -28593.0676, -5996.3414, -3211.6816] \tag{10.17}$$
$$\beta_c = [-93730.1448, 313067.8210, -345133.6546] \tag{10.18}$$

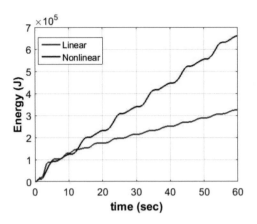

Figure 10.3: Extracted energy for case 1.

Simulations are conducted for 60 s. For this illustrative case, the extracted energy from the linear system (assuming no nonlinearities in the system) is 3.247588×10^5 Joules while the extracted energy from the nonlinear system is 6.615232×10^5 Joules, in 60 s as shown in **Fig.** 10.3. The nonlinear system produces about 2.0 times the energy harvested by the linear system. **Figure** 10.4 shows the power extracted by each system. As expected, the linear system does not need any reactive power since the control is just a linear damping. The nonlinear system in this case needs reactive power, the amount of which is very small though compared to the harvested power, as shown in **Fig.** 10.4 where a zoom is made on a small part to show the level of reactive power. Note that the

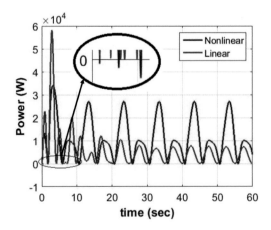

Figure 10.4: No reactive power in both linear and nonlinear controls.

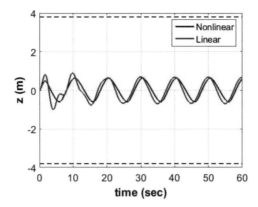

Figure 10.5: Displacements of both linear and nonlinear systems are in the same range.

amount of reactive power dictates the size of the energy storage needed by the control system; and the control designs that require less reactive power have the advantage of lower cost and less complexity compared to those control strategies that require bigger energy storage. **Figure** 10.5 shows the displacement for both systems, where the dash horizontal line represents the z_s limit beyond which the buoy is either totally in the air or totally submerged. After an initial transition period, the displacements of both WECs have about the same amplitude and frequency; the same can be said about the velocities. This result highlights the significant advantage that can be obtained when designing both the shape and the control simultaneously. When designing for higher harvested energy, the control design usually exploits the system nonlinearities and results in higher amplitude motions as compared to the linear motion. In this case study, however, the nonlinear system has about the same motion amplitude as the linear one. The reason is due to the nonlinear system force. The latter seem to cancel the effect of the nonlinear control force on the buoy motion. Yet, in computing the harvested power, only the control force is multiplied by the buoy velocity; the system force does not cancel the nonlinear control force in the power calculation. If the reference amplitude of excitation force is increased to 1319900 N, the amount of harvested energy increases to about 2.7 that of the linear system; with minimal need for reactive power and a small motion amplitude.

It is required to find the optimal number of nonlinear control and system coefficients, as well as their values, such that the objective function (extracted energy) is maximized. The case study presented in **Section** 10.2 is here optimized to maximize the harvested energy. A population of 20 members along with only 5 generations are used in optimization. The number of elite members is 2. A saturation limit is assumed on the nonlinear control force $\tilde{u}_c$; this saturation limit is equal to $1.1 \times max(f_{ex})$. Also solutions that result in very high motion amplitude are excluded. The maximum possible value for each of N_s and N_c is selected to be 6, while the maximum possible value for each of M_s and M_c is selected to be 4. This means that the chromosome is structured such that it has 6 genes for α_s, 6 genes for α_c, 4 genes for β_s, and 4 genes for β_c. The total number of genes is then 20; the lower and upper bounds of these design variables are listed in **Table** 10.2. The input data for the optimizer are summarized in **Table** 10.3.

Table 10.2: Case 1: bounds on design variables.

Design Variable	Value
$\alpha_{c_{min}}$	$[0, -100 \times (m/(rT_w^2)), -100 \times (m/(r^2 T_w^2)), -100 \times (m/(r^3 T_w^2)), -100 \times (m/(r^4 T_w^2)), -100 \times (m/(r^5 T_w^2))]$
$\alpha_{s_{min}}$	$[0, -100 \times (m/(rT_w^2)), -100 \times (m/(r^2 T_w^2)), -100 \times (m/(r^3 T_w^2)), -100 \times (m/(r^4 T_w^2)), -100 \times (m/(r^5 T_w^2))]$
$\alpha_{c_{max}}$	$[0, 100 \times (m/(rT_w^2)), 100 \times (m/(r^2 T_w^2)), 100 \times (m/(r^3 T_w^2)), 100 \times (m/(r^4 T_w^2)), 100 \times (m/(r^5 T_w^2))]$
$\alpha_{s_{max}}$	$[0, 100 \times (m/(rT_w^2)), 100 \times (m/(r^2 T_w^2)), 100 \times (m/(r^3 T_w^2)), 100 \times (m/(r^4 T_w^2)), 100 \times (m/(r^5 T_w^2))]$
$\beta_{c_{min}}$	$[-20 \times (m/(T_w)), -20 \times (m/r), -20 \times (mT_w/r^2), -20 \times (mT_w^2/r^3)]$
$\beta_{s_{min}}$	$[-20 \times (m/(T_w)), -20 \times (m/r), -20 \times (mT_w/r^2), -20 \times (mT_w^2/r^3)]$
$\beta_{c_{max}}$	$[20 \times (m/(T_w)), 20 \times (m/r), 20 \times (mT_w/r^2), 20 \times (mT_w^2/r^3)]$
$\beta_{s_{max}}$	$[0, 0, 0, 0]$

Table 10.3: Case 1: Input data to the optimizer.

Parameter	Value
Population Size	20
Number of Generations	5
Elite Count	2
Maximum Control $\tilde{u}_{c,max}$	$1.1 \times max(f_{ex})$
Maximum N_s	6
Maximum N_c	6
Maximum M_s	4
Maximum M_c	4

Some of the design variables are hidden in each chromosome. Different chromosomes, in general, have different hidden genes; which means that different solutions may have different nonlinear terms in the system and control forces.

The results of the optimization are presented in **Table** 10.4; where the best found solution has 3 hidden genes in N_c, 4 hidden genes in N_s, 4 hidden genes in M_s, and no hidden genes in M_c. The value of each of the variables represented by hidden genes in the optimal solution is listed as '0' in **Table** 10.4. The optimal values of the active (non hidden) variables are also listed. With these optimized variables, the energy harvested by the nonlinear system is 1.591022×10^6 Joules. This corresponds to about 4.9 times the energy harvested by the linear system simulated in **Section** 10.2. **Figure** 10.6 shows the performance of this optimized system where the harvested power is shown to be significantly higher than that of the linear system while still needing only a small amount of reactive power.

Table 10.4: Optimization Results for Case study 1.

Design Variable	Value
α_c	$[0, 35559.9334, -12073.5531, 0, 0, 0]$
α_s	$[0, 37675.7192, 0, 0, 0, 0]$
β_c	$[-297088.6252, 858572.1382, -735890.0019, -1386250.3445]$
β_s	$[-10556.6641, 0, 0, 0]$

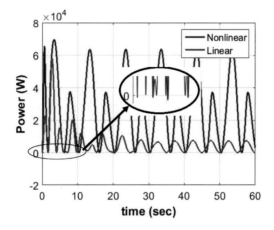

Figure 10.6: Extracted power for optimized Case 1.

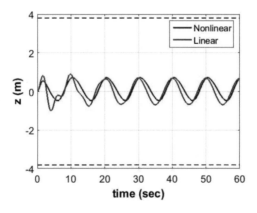

Figure 10.7: Buoy position for optimized Case 1.

The motion of the nonlinear system is in the same order of magnitude as that of the linear one, as depicted in **Fig.** 10.7.

10.3 WECs in Irregular Waves

The more realistic case is when the WEC is in an irregular wave. In this section we assume a Bretschneider wave spectrum. The main difference between this case and the previous regular wave case is in the calculations of the radiation and excitation forces. Here, the radiation force is frequency dependant and is represented as added mass and added damping. The excitation force is also frequency dependant. The equation of motion of the buoy in such case is:

$$\bar{m}\ddot{z} = f_{ex} + u_l + \tilde{u}_c + \tilde{f}_s + f_r + f_B \tag{10.19}$$

where $\bar{m}$ is the mass in addition to the added mass at infinite frequency, f_B is the linear buoyancy force, f_r is the linear radiation damping force, and f_{ex} is the linear excitation force. All the nonlinear system forces are collected in the term $\tilde{f}_s$. The nonlinear control is $\tilde{u}_c$. The linear radiation damping force f_r can be computed using a state space model of the form:

$$\dot{\vec{x}}_r = A_r\vec{x}_r + B_r\dot{z}, \tag{10.20}$$
$$f_r = C_r\vec{x}_r \tag{10.21}$$

where, $\vec{x}_r$ represents the radiation states. The matrices A_r, B_r, and C_r are calculated as functions of the added masses and added damping data. Note that when $|z| > z_s$ and $z < 0$, then $f_r = 0$ and $\bar{m} = m$. The buoyancy force on the buoy, f_B, can be calculated as:

$$f_B = \begin{cases} mg - \rho g V_{b_s} & \text{if } |z| < z_s \\ mg - \rho g V_b & \text{if } |z| > z_s, z > 0 \\ mg & \text{if } |z| > z_s, z < 0 \end{cases} \tag{10.22}$$

where V_{b_s} is the submerged volume, and V_b is the total buoy volume. The linear excitation force, f_{ex}, is:

$$f_{ex} = \sum_{n=1}^{N} a_n f_n \exp\left(-i\left(\omega_n t - \phi_n\right)\right) \tag{10.23}$$

where N is the number of frequencies used to realize the wave, a_n are the wave coefficients, f_n are excitation force coefficients, ω_n are frequencies, and ϕ_n are phase shifts.

Consider the same system presented in **Section** 10.2, but this time assuming a Bretschneider wave spectrum with a peak period of 10 sec, and a wave amplitude of 0.7 m. Let the frequency range be $\omega = [0.01 : 0.037 : 7]$ rad/sec. The initial phase shifts are randomly selected in the range $[-\pi, \pi]$.

The nonlinear system and control coefficients are optimized using HGGA; the upper and lower bounds for the design variables are shown in **Table** 10.5. The obtained results are presented in **Table** 10.6. The total energy harvested

Table 10.5: Lower and upper bounds of design variables.

Design Variable	Value
$\alpha_{c_{min}}$	$[0, -10^6, -10^6, -10^6, -10^6, -10^6]$
$\alpha_{s_{min}}$	$[0, -10^6, -10^6, -10^6, -10^6, -10^6]$
$\alpha_{c_{max}}$	$[0, 10^6, 10^6, 10^6, 10^6, 10^6]$
$\alpha_{s_{max}}$	$[0, 10^6, 10^6, 10^6, 10^6, 10^6]$
$\beta_{c_{min}}$	$[-2 \times 10^6, -2 \times 10^6, -2 \times 10^6, -2 \times 10^6]$
$\beta_{s_{min}}$	$[-2 \times 10^6, -2 \times 10^6, -2 \times 10^6, -2 \times 10^6]$
$\beta_{c_{max}}$	$[2 \times 10^6, 2 \times 10^6, 2 \times 10^6, 2 \times 10^6]$
$\beta_{s_{max}}$	$[0, 0, 0, 0]$

Table 10.6: Optimization results for the Bretschneider wave case.

Design Variable	Value
α_c	$[0, -508230.0876, 389006.9833]$
α_s	$[0, -190490.8616]$
β_c	$[1821835.7346, -1630799.6246]$
β_s	$[-849431.0745]$

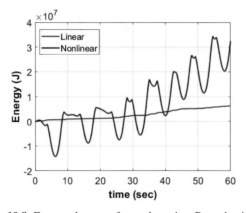

Figure 10.8: Extracted energy from a buoy in a Bretschneider wave.

in 60 seconds of simulation is 6.296549×10^6 Joules from the baseline linear system. The total energy harvested from the nonlinear system is 3.215181×10^7 Joules, which is about 5 times that harvested from the linear buoy, as shown in **Fig.** 10.8. Significant reactive power is needed in this case. The nonlinear system motion is significantly higher than that of the linear system as shown in **Fig.** 10.9. **Figure** 10.10 shows the buoy velocity over time and **Fig.** 10.11 shows the control force on the buoy, for both the linear and nonlinear cases. Note that in this simulation, a saturation limit is assumed on the control force of 1.25×10^6 N.

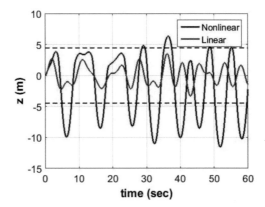

Figure 10.9: Buoy position in a Bretschneider wave.

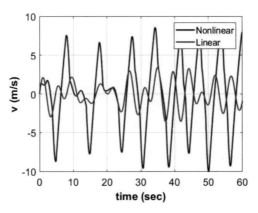

Figure 10.10: Buoy velocity in a Bretschneider wave.

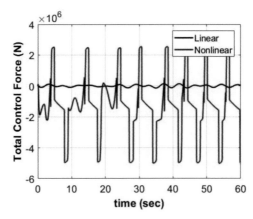

Figure 10.11: The control force for the buoy in a Bretschneider wave.

10.3.1   Simultaneous Optimization of Shape and Control

Another case study is here presented. Here we assume the buoy shape is divided into multiple lateral sections. The classical cylindrical buoy consists of one section, of a cylindrical shape, with a constant cross section area. Here, we assume multiple sections, each section may have a different shape. For example, a buoy may have three sections, the top one is cylindrical, the middle one is spherical, and the bottom one is conical. The discussion presented in **Section** 2.3 shows that it is possible to compute the nonlinear Froude Krylov force for each section separately, and combine them. **Section** 2.3 also shows how to compute the nonlinear Froude Krylov force for different shapes. Here is this section, we let the optimizer search for the optimal shapes of all sections (section elements), along with the optimal control coefficients. Genetic algorithms are used for optimization.

For this test, the GA searches also for the optimal number of different sections. Here the number of section elements is constrained to be an integer between 3 and 5. A range of mass from 380kg to 500 kg was selected for the buoy. The shape and control coefficients were optimized to maximize the ratio of the steady state power over the mass of the buoy.

Simulation results are shown in **Figs.** 10.12, 10.13, 10.14, and 10.15. This test case considers only axisymmetric heaving buoy design, so only right half of the cross-session on x–z plane is shown in **Fig.** 10.12(a). The motion of the nonlinear buoy is almost of the same amplitude as that of the linear one as shown in **Fig.** 10.12(b).

The harvested energy using the nonlinear buoy is higher than that of the linear buoy as shown in **Fig.** 10.13. The new shape has a mass of 401 kg, similar to the 399 kg baseline cylindrical buoy. **Figure** 10.16 shows a three-dimensional render for the optimized buoy shape as compared to a typical cylindrical buoy. The energy ratio of the new shape in steady state was 18 W/kg. However, the power quality was not good as the ratio between the mean and maximum power was smaller then the baseline design.

One advantage of this new design for the buoy is that the required control force is not significantly higher than that of a cylindrical buoy. **Figure** 10.3.1 shows a comparison between the control of this new buoy design and a cylindrical buoy. In this simulation, a hard constraint is imposed on the nonlinear WEC control. If this hard constraint is removed, it is observed that there is only very few spikes of hih control force, while most of the time the control force is within the desired bounds. That is why we imposed this hard constraint, without significant impact on the system performance, as evident from the simulation results. **Figure** 10.3.1 is provided to explain how the excitation force on the new buoy change over time, in relation to the buoy position. Specifically, when the buoy is mostly in the air, the excitation force is almost zero. When the buoy is mostly under water, the excitation forces reaches its maximum value.

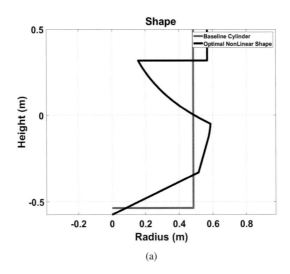

(a)

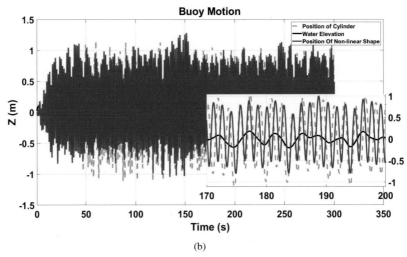

(b)

Figure 10.12: Comparison between the optimal solution and the baseline cylindrical Buoy shape, along with the displacement of both types of buoys in a Bretschnieder wave.

10.4 Discussion

This chapter presents a study that attempts to optimize the nonlinear control along with the system nonlinearities to maximize the harvested energy. It is assumed that these nonlinear system forces can be generated during the system design process (such as the buoy shape design) and/or varying the buoy shape in realtime. Note that it is possible to extract the coefficients of the system nonlinear force ($\tilde{f}_s$) using CFD simulations. A nonlinear least squares fitting algorithm,

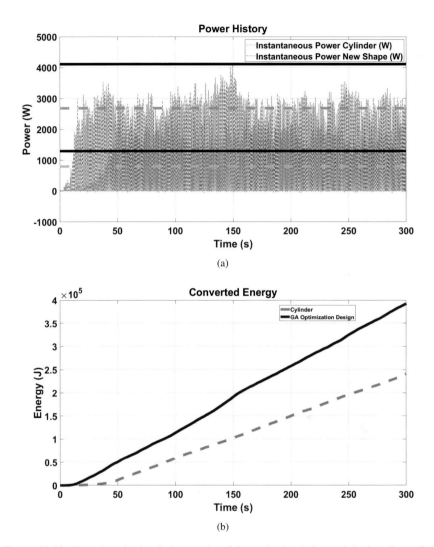

Figure 10.13: Time domain simulation results of the optimal solution and the baseline cylindrical buoy, in terms of instantaneous power, mean power, maximum power, and total converted energy. Solid horizontal lines in **Fig.** 10.13(a) represent the maximum power and the average power of the optimal non-linear shape design, dashed horizontal lines in **Fig.** 10.13(a) represent the maximum power and the average power of the baseline shape.

for example, can be used to fit the expansion in Eq. (10.5) to the nonlinear forces obtained from the CFD simulations. This process can be used iteratively for designing a buoy shape that generates approximate prescribed nonlinear forces. This process, however, is not addressed in this chapter. Consider the case when it is not possible to find a buoy shape that would generate a prescribed (optimized)

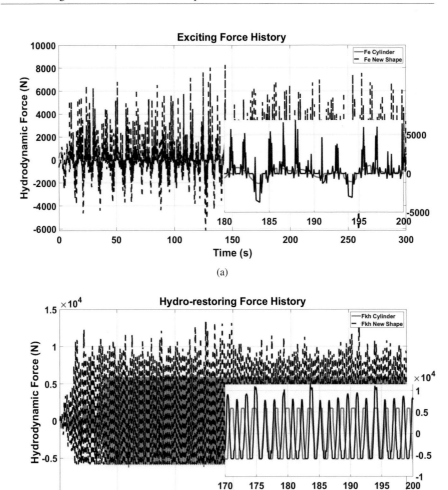

(a)

(b)

Figure 10.14: Different hydrodynamic force history for the optimal solution without control constraint and the baseline WEC in the time domain simulation.

nonlinear force $\tilde{f}_s$. In such case, it is possible to modify the nonlinear control terms to compensate for part or all of the unattainable nonlinear $\tilde{f}_s$ force. For instance, consider the case study presented in **Section** 10.2 and assume that it is not possible to design a buoy shape that can provide the optimized force $\tilde{f}_s$. In such case it is possible to assume the buoy shape to be cylindrical, for instance, and set $\tilde{f}_s = 0$. The HGGA is then used to optimize a nonlinear control force $\tilde{u}_c$ (Eq. (10.4)). This case is carried out and the results are shown in **Fig.** 10.19. The

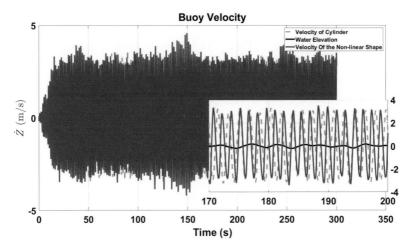

Figure 10.15: Different velocity history for the optimal solution without control constraint and the baseline WEC in the time domain simulation.

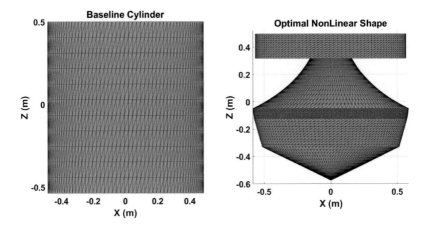

Figure 10.16: The optimized buoy shape compared to a typical cylindrical buoy.

harvested energy in this case is $4.298156e \times 10^5$ Joules, which is about 1.32 times the harvested energy from the linear system in **Section** 10.2. Compare this energy amplification factor to the factor of 4.9 obtained in **Section** 10.2 when both $\tilde{f}_s$ and $\tilde{u}_c$ were optimized. This result supports the claim that a design optimization process that involves both the control and system nonlinearities might lead to more efficient energy conversion, compared to the optimization of the control only for a given system.

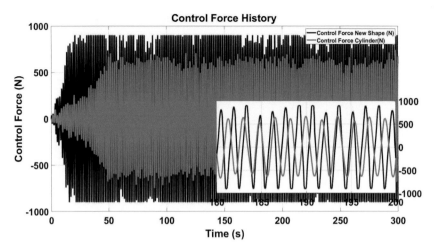

Figure 10.17: The control force of the new nonlinear buoy is comparable to that of a classical cylindrical buoy.

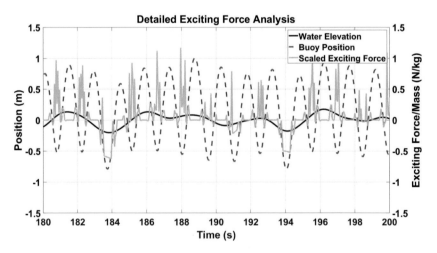

Figure 10.18: Scaled excitation force over time along with the buoy position shows that the excitation force vanishes when the buoy is mostly in the air, and it goes to a maximum when the buoy is mostly under water.

Section 2.3 presents the dynamic model adopted in the case of large motion, which results in nonlinear effects. It is possible to eliminate these effects by limiting the motion of the buoy, using movement limiting or end-stop mechanisms for instance. One of the goals of this study, however, is to investigate the possibility of increasing the harvested energy when these nonlinearities are exploited, as opposed to being avoided. Hence, the general approach in this chapter is not to avoid nonlinear forces; but rather an attempt to model these nonlinearities and optimize a control taking into consideration these nonlinear effects.

It is also noted that if the system has non-optimizable nonlinear forces, then these forces can be added to the equation of motion in Eq. 10.1. Such case does not affect the generality of the proposed method in this chapter, since the optimizer will search for the optimal parameters in the optimizable forces in the presence of the non-optimizable ones.

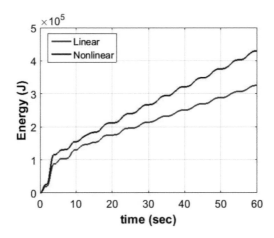

Figure 10.19: Harvested energy – no system nonlinearities.

Finally, the optimization process described in this chapter implements HGGA which might not be suitable for realtime implementation. The HGGA is used in this chapter for the purpose of design. Once the buoy shape is optimized, the obtained values for the α_s and β_s coefficients become constants that do not change during operation. Regarding the control, the coefficients α_c and β_c can be optimized off line for a range of sea states, and stored in a look-up table. During operation, this look-up table can be used to select or interpolate the appropriate values for α_c and β_c. Moreover, if a further tuning is needed during operation, this look-up table can be used to provide a good initial guess for a faster optimization algorithm. This detail is not addressed in this chapter.

Bibliography

[1] Abdelkhalik, O. 2011. Multi-gravity-assist trajectories optimization: Comparison between the hidden genes and the dynamic-size multiple populations genetic algorithms. *In*: Proceedings of the 2011 AAS/AIAA Astrodynamics Specialist Conference, Girdwood, Alaska, USA.

[2] Abdelkhalik, O. 2013. Autonomous planning of multigravity-assist trajectories with deep space maneuvers using a differential evolution approach. International Journal of Aerospace Engineering 2013(145369). doi:10.1155/2013/145369.

[3] Abdelkhalik, O. 2013. Hidden genes genetic optimization for variable-size design space problems. Journal of Optimization Theory and Applications 156(2): 450–468.

[4] Abdelkhalik, O. and S. Darani. 2016. Hidden genes genetic algorithms for systems architecture optimization. *In*: Proceedings of the Genetic and Evolutionary Computation Conference 2016, GECCO'16, pp. 629–636, New York, NY, USA. ACM.

[5] Abdelkhalik, O. and S. Darani. 2018. Evolving hidden genes in genetic algorithms for systems architecture optimization. ASME Journal of Dynamic Systems, Measurement and Control 140(10): 101015-1–101015-11, October 2018.

[6] Abdelkhalik, O. and S. Darani. 2018. Optimization of nonlinear wave energy converters. Ocean Engineering 162: 187–195.

[7] Abdelkhalik, O. and A. Gad. 2011. Optimization of space orbits design for earth orbiting missions. Acta Astronautica, Elsevier 68: 1307–1317, April–May 2011.

[8] Abdelkhalik, O. and A. Gad. 2012. Dynamic-size multiple populations genetic algorithm for multigravity-assist trajectories optimization. AIAA Journal of Guidance, Control, and Dynamics 35(2): 520–529, March–April 2012.

[9] Abdelkhalik, O. and D. Mortari. 2006. Orbit design for ground surveillance using genetic algorithms. Journal of Guidance, Control, and Dynamics 29(5): 1231–1235, September–October 2006.

[10] Abdelkhalik, O. and D. Mortari. 2007. On the n-impulse orbit transfer using genetic algorithms. Journal of Spacecraft and Rockets 44(2): 456–460, March–April 2007.

[11] Abdelkhalik, O., R. Robinett, G. Bacelli, R. Coe, D. Bull, D. Wilson and U. Korde. 2015. Control optimization of wave energy converters using a shape-based approach. *In*: ASME Power & Energy, San Diego, CA, June 2015. ASME.

[12] Abdelkhalik, O., R. Robinett, S. Zou, G. Bacelli, R. Coe, D. Bull, D. Wilson and U. Korde. 2016. On the control design of wave energy converters with wave prediction. Journal of Ocean Engineering and Marine Energy 2(4): 473–483.

[13] Abdelkhalik, O., J. Song, R. Robinett, G. Bacelli, D. Wilson and U. Korde. 2016. Feedback control of wave energy converters. *In*: Asian Wave and Tidal Energy Conference (AWTEC 2016), pp. 258–261, Marina Bay Sands, Singabore, October 2016.

[14] Abdelkhalik, O., S. Zou, R. Robinett, G. Bacelli, D. Wilson. 2017. Estimation of excitation forces for wave energy converters control using pressure measurements. International Journal of Control 90(8): 1793–1805.

[15] Abilleira, F. 2007. Broken-plane maneuver applications for earth to mars trajectories. *In*: 20th International Symposium on Space Flight Dynamics, September 2007.

[16] Abraham, E. 2013. Optimal Control and Robust Estimation for Ocean Wave Energy Converters. Phd thesis, Department of Aeronautics, Imperial College London.

[17] Abu-Lebdeh, G. and R.F. Benekohal. 2000. Signal coordination and arterial capacity in oversaturated conditions. Journal of the Transportation Research Board, TRB, National Research Council, Washington, DC 1727: 68–76.

[18] Alexandru Agapie. 2005. Genetic algorithms: Minimal conditions for convergence. *In*: Proceedings of the 2005 European Conference on Artificial Evolution, pp. 181–193, Heidelberg, Berlin, Germany.

[19] Allison, J.T., A. Khetan and D. Lohan. 2013. Managing variable-dimension structural optimization problems using generative algorithms. *In*: 10th World Congress on Structural and Multidisciplinary Optimization, Orlando, Florida, USA, May 19–24.

[20] Bacelli, G., R. Genest and J.V. Ringwood. 2015. Nonlinear control of flap-type wave energy converter with a non-ideal power take-off system. Annual Reviews in Control 40: 116–126.

[21] Bacelli, G., J.V. Ringwood and J.-C. Gilloteaux. 2011. A control system for a self-reacting point absorber wave energy converter subject to constraints.*In*: Proceedings of the 18th IFAC World Congress, pp. 11387–11392, Milano, Italy.

[22] Bacelli, G. 2014. Optimal control of wave energy converters. PhD, National University of Ireland, Maynooth, Maynooth, Ireland.

[23] Bandyopadhyay, S. and S.K. Pal. 2001. Pixel classification using variable string genetic algorithms with chromosome differentiation. IEEE Transactions on Geoscience and Remote Sensing 39(2): 303–308, February 2001.

[24] Battin, R.H. 1987. An Introduction to the Mathematics and Methods of Astrodynamics. AIAA.

[25] Beasley, D., D.R. Bull and R.R. Martin. 1993. A sequential niche technique for multimodal function optimization. Evolutionary Computation 1(2): 101–125, Summer 1993.

[26] Boyd, S. and L. Vandenberghe. 2004. Convex Optimization. Cambridge University Press, New York, NY, USA.

[27] Chang, T.-H. and J.T. Lin. 2000. Optimal signal timing for an oversaturated intersection. Transportation Research Part B: Methodological 34(6): 471–491.

[28] Cretel, J.A.M., G. Lightbody, G.P. Thomas and A.W. Lewis. 2011. Maximisation of energy capture by a wave-energy point absorber using model predictive control. *In*: IFAC Proceedings 44.1: 3714–3721.

[29] Cummins, W.E. 1962. The Impulse Response Function and Ship Motions. David W. Taylor Model Basin Report.

[30] Vinkò, T. and D. Izzo. 2010. GTOP act trajectory database. Technical report, European Space Agency[online database]. http://www.esa.int/gsp/ACT/inf/op/globopt.htm.

[31] Davis, T.E. and J.C. Principe. 1993. A markov chain framework for the simple genetic algorithm. Evolutionary Computation 1: 269–288.

[32] Derrick, J., S. Garcia, D. Molina and F. Herrera. 2011. A practical tutorial on the use of nonparametric statistical tests as a methodology for comparing evolutionary and swarm intelligence algorithms. Journal of Swarm and Evolutionary Computation 1.

[33] Dorigo, M., V. Maniezzo and A. Colorni. 1996. The ant system: Optimization by a colony of cooperating agents. IEEE Transactions on Systems, Man, and Cybernetics-Part B 26: 29–41.

[34] Drozdek, A. 2001. Data Structures and Algorithms in C++. Brook/Cole, Pacific Grove, CA, second edition.

[35] Eggermont, J., A.E. Eiben and J.I. van Hemert. 1999. A comparison of genetic programming variants for data classification. *In*: Proceedings on the Third Symposium on Intelligent Data Analysis, (IDA-99) LNCS 1642.

[36] Crossley, W.A., T.A. Ely and E.A. Williams. 2009. Satellite constellation design for zonal coverage using genetic algorithms. Journal of the Astronautical Sciences 47(3-4): 207–228, July–December 1999.

[37] Englander, J., B. Conway and T. Williams. 2012. Automated mission planning via evolutionary algorithms. AIAA Journal of Guidance, Control, and Dynamics 35(6): 1878–1887.

[38] Englander, J.A., B.A. Conway and B.J. Wall. 2009. Optimal strategies found using genetic algorithms for deflecting hazardous near-earth objects. *In*: CEC'09: Proceedings of the Eleventh Conference on Congress on Evolutionary Computation, pp. 2309–2315, Piscataway, NJ, USA. IEEE Press.

[39] Lim, L.F., L.A. McFadden, A.R. Rhoden, K.S. Noll, J.A. Englander and M.A. Vavrina. 2017. Trajectory optimization for missions to small bodies with a focus on scientific merit. Comput. Sci. Eng. 19(4): 18–28.

[40] ESA. 2009. Jupiter ganymede orbiter - esa contribution to the europa jupiter system mission. Assessment Study Report ESA-SRE(2008)2, European Space Agency, March 2009.

[41] European Space Agency. 2009. GTOP act trajectory database. http://www.esa.int/gsp/ACT/inf/op/globopt.htm, July 2009.

[42] Faedo, N., S. Olaya and J.V. Ringwood. 2017. Optimal control, mpc and mpc-like algorithms for wave energy systems: An overview. IFAC Journal of Systems and Control 1: 37–56.

[43] Falkenauer, E. 1991. A genetic algorithm for grouping. pp. 198–206. *In*: R. Gutierrez and M.J. Valerrama (eds.). The Fifth International Symposiom on Applied Stochastic Models and Data Analysis, Granada, Spain, 23–26 April 1991. World Scientific Publishing Co., Singapore.

[44] Falkenauer, E. 1996. A hybrid grouping genetic algorithm for bin packing. Journal of Heuristics, Springer Netherlands 2(1): 5–30, June 1996.

[45] Falkenauer, E. 1998. Genetic Algorithms and Grouping Problems. Wiley.

[46] Falnes, J. 2002. Ocean Waves and Oscillating Systems–Linear Interactions Including Wave-Energy Extraction. Cambridge University Press.

[47] Falnes, J. 2007. A review of wave-energy extraction. Marine Structures 20(4): 185–201, October 2007.

[48] Fusco, F. and J. Ringwood. 2014. Hierarchical robust control of oscillating wave energy converters with uncertain dynamics. Sustainable Energy, IEEE Transactions on 5(3): 958–966, July 2014.

[49] Gad, A. and O. Abdelkhalik. 2011. Hidden genes genetic algorithm for multi-gravity-assist trajectories optimization. AIAA Journal of Spacecraft and Rockets 48(4): 629–641, July–August 2011.

[50] Gad, A., N. Addanki and O. Abdelkhalik. 2010. N-impulses interplanetary orbit transfer using genetic algorithm with application to Mars mission. *In*: 20th AAS/AIAA Space Flight Mechanics Meeting, Number AAS Paper 10–167, San Diego, CA, USA.

[51] Gauss, C.F. and G.W. Stewart. 1995. 1. Theory of the Combination of Observations Least Subject to Errors: Part One, pp. 2–47.

[52] Gauss, C.F. and G.W. Stewart. 1995. 2. Theory of the Combination of Observations Least Subject to Errors: Part Two, pp. 50–97.

[53] Gazis, D.C. 1964. Comparing ant colony optimization and genetic algorithm. Operations Research 12: 815–831.

[54] Gazis, D.C. and R.B. Potts. 1965. The oversaturated intersection. *In*: Proceedings of the Second International Symposium on the Theory of Road Traffic Flow, pp. 221–237, Paris. Organization for Economic Cooperation and Development.

[55] Gentile, L., G. Filippi, E. Minisci, T. Bartz-Beielstein and M. Vasile. 2020. Preliminary spacecraft design by means of structured-chromosome genetic algorithms. *In*: IEEE World Congress on Computational Intelligence (IEEE WCCI), Glasgow, United Kingdom, 07 2020.

[56] Gentile, L., C. Greco, E. Minisci, T. Bartz-Beielstein and M. Vasile. 2019. Structured-chromosome ga optimisation for satellite tracking. *In*: GECCO'19 Proceedings of the Genetic and Evolutionary Computation Conference Companion, pp. 1955–1963. ACM, Prague, Czech Republic, September 2019.

[57] Gentile, L., E. Morales, D. Quagliarella, E. Minisci, T. Bartz-Beielstein and R. Tognaccini. 2020. High-lift devices topology optimisation using structured-chromosome genetic algorithm. *In*: IEEE World Congress on Computational Intelligence (IEEE WCCI), Glasgow, United Kingdom, 07 2020.

[58] Giorgi, G. and J.V. Ringwood. 2016. Computationally efficient nonlinear froude–krylov force calculations for heaving axisymmetric wave energy point absorbers. Journal of Ocean Engineering and Marine Energy, 1–13.

[59] Giorgi, G. and J.V. Ringwood. 2018. Analytical formulation of nonlinear froude-krylov forces for surging-heaving-pitching point absorbers. *In*: ASME 2018 37th International Conference on Ocean, Offshore and Arctic Engineering, pp. V010T09A036–V010T09A036. American Society of Mechanical Engineers.

[60] Giorgi, G., M.P. Retes and J. Ringwood. 2016. Nonlinear hydrodynamic models for heaving buoy wave energy converters. *In*: Asian Wave and Tidal Energy Conference (AWTEC 2016), pp. 144–153, Marina Bay Sands, Singabore, October 2016.

[61] Girianna, M. and R.F. Benekohal. 2004. Using genetic algorithms to design signal coordination for oversaturated networks. Journal of Intelligent Transportation Systems (8): 117–129.

[62] Goldberg, D.E. 1989. Genetic Algorithms in Search, Optimization and Machine Learning. Addison-Wesley Longman Publishing Co., Inc., Boston, MA, USA, 1st Edition.

[63] Guo, X., W. Zhang and W. Zhong. 2014. Topology optimization based on moving deformable components: A new computational framework. Journal of Applied Mechanics, 81.

[64] Hal's, J., J. Falnes and T. Moan. 2011. Constrained optimal control of a heaving buoy wave-energy converter. Journal of Offshore Mechanics and Arctic Engineering 133(1): 1–15.

[65] Becerra, V., D. Myatt, S. Nasuto, D. Izzo and J. Bishop. 2007. Search space pruning and global optimisation of multiple gravity assist spacecraft trajectories. Journal of Global Optimisation 38(2): 283–296.

[66] Izzo, D., V.M. Becerra, D.R. Myatt, S.J. Nasuto and J.M. Bishop. 2007. Search space pruning and global optimisation of multiple gravity assist spacecraft trajectories. Journal of Global Optimisation 38(2): 283–296, June 2007.

[67] Jafarpour, B. and D.B. McLaughlin. 2007. History matching with an ensamble kalman filter and discrete cosine parameterization. SPE 108761, November 2007.

[68] Izzo, D. and J. Olympio. 2009. Designing optimal multi-gravity-assist trajectories with free number of impulses. *In*: 21st International Symposium on Space Flight Dynamics, Toulouse, France, September 2009.

[69] Kennedy, J. and R.C. Eberhart. 1995. Particle swarm optimization. *In*: Proceedings of the 1995 IEEE International Conference on Neural Networks, volume 4, pp. 1942–1948, Perth, Australia, IEEE Service Center, Piscataway, NJ.

[70] Kim, H.-D., O.-C. Jung and H. Bang. 2007. A computational approach to reduce the revisit time using a genetic algorithm. International Conference on Control, Automation and Systems, pp. 184–189, October 2007.

[71] Kim, Y.H. and D.B. Spencer. 2002. Optimal spacecraft rendezvous using genetic algorithms. Journal of Spacecraft and Rockets 39(6), November–December 2002.

[72] Korde, U.A., M.P. Schoen and F. Lin. 2001. Time domain control of a single mode wave energy device. *In*: Proceedings of the Eleventh International Offshore and Polar Engineering Conference, pp. 555–560, Stavanger, Norway.

[73] Korde, U.A. 1998. On control approaches for efficient primary energy conversion in irregular waves. *In*: OCEANS'98 Conference Proceedings, volume 3, pp. 1427–1431, September 1998.

[74] Ku, S. and B. Lee. 2001. A set-oriented genetic algorithm and the knapsack problem. *In*: Evolutionary Computation, 2001. Proceedings of the 2001 Congress on 1: 650–654.

[75] Li, Guang. Nonlinear model predictive control of a wave energy converter based on differential flatness parameterisation. International Journal of Control 89(0): 1–10.

[76] Li, G. and M.R. Belmont. 2014. Model predictive control of sea wave energy converters Part I: A convex approach for the case of a single device. Renewable Energy 69(0): 453–463.

[77] Li, G., G. Weiss, M. Mueller, S. Townley and M.R. Belmont. 2012. Wave energy converter control by wave prediction and dynamic programming. Renewable Energy 48(0): 392–403.

[78] Lieberman, E. and J. Chang. 2005. Optimizing traffic signal timing through network decomposition. Transportation Research Record: Journal of the Transportation Research Board 1925(1): 167–175.

[79] Luo, Y. and G.T. Schuster. 1999. Wave-equation Travel Time Inversion. Geophysics, May 1999.

[80] Maher, M. 2008. The optimization of signal settings on a signalized round-about using the cross-entropy method. Computer-Aided Civil and Infrastructure Engineering 23(2): 76–85.

[81] Maulik, U. and Indrajit Saha. 2009. Modified differential evolution based fuzzy clustering for pixel classification in remote sensing imagery. Pattern Recognition 42(9): 2135–2149.

[82] Mérigaud, A., J.-C. Gilloteaux and J.V. Ringwood. 2012. A nonlinear extension for linear boundary element methods in wave energy device modelling. Volume 4: Offshore Geotechnics; Ronald W. Yeung Honoring Symposium on Offshore and Ship Hydrodynamics, January 2012.

[83] Napier, S.W. and J.W. McMahon. 2018. A novel multi-spacecraft interplanetary global trajectory optimization transcription. AAS 18–401 of the 2018 AAS/AIAA Astrodynamics Specialist Conference, August 19–23, 2018.

[84] Navagh, J. 1993. Optimizing interplanetary trajectories with deep space maneuvers. Nasa contractor report 4546, NASA Langley Research Center, Hampton, VA.

[85] Nayfeh, A.H. and D.T. Mook. 2008. Nonlinear Oscillations. Wiley Classics Library. Wiley.

[86] Nix, A.E. and M.D. Vose. 1992. Modeling genetic algorithms with markov chains. Annals of Mathematics and Artificial Intelligence 5: 79–88.

[87] Nyew, H.M., O. Abdelkhalik and N. Onder. 2015. Structured chromosome evolutionary algorithms for variable-size autonomous interplanetary trajectory planning optimization. Journal of Aerospace Information Systems 12(3): 314–328.

[88] Oetinger, D., M.E. Magaña and O. Sawodny. 2004. Decentralized model predictive control for wave energy converter arrays. Sustainable Energy, IEEE Transactions on 5(4): 1099–1107, October 2014.

[89] Olds, A.D., C.A. Kluever and M.C. Cupples. 2007. Interplanetary mission design using differential evolution. Journal of Spacecraft and Rockets 44(5): 1060–1070.

[90] Olympio, J.T. 2009. Designing optimal multi-gravity-assist trajectories with free number of impulses. *In*: 21st International Symposium on Space Flight Dynamics, Toulouse, France.

[91] Olympio, J.T. and J.P. Marmorat. 2007. Global trajectory optimisation: Can we prune the solution space when considering deep space maneuvers? Final report. European Space Agency, September 2007.

[92] Park, J.-M., J.-G. Park, C.-H. Lee and M.-S. Han. 1993. Robust and efficient genetic crossover operator: homologous recombination. *In*: Neural Networks, 1993. IJCNN'93-Nagoya. Proceedings of 1993 International Joint Conference on, volume 3, pp. 2975–2978, October 1993.

[93] Perez, T. and T.I. Fossen. 2009. A MATLAB toolbox for parametric identification of radiation-force models of ships and offshore structures. Modeling, Identification and Control: A Norwegian Research Bulletin 30(1): 1–15.

[94] Price, K., Rainer M. Storn and J. A. Lampinen. 2005. Differential evolution: A practical approach to global optimization. Natural Computing Series. Springer-Verlag Berlin Heidelberg.

[95] Putha, R., L. Quadrifoglio and E.M. Zechman. 2012. Comparing ant colony optimization and genetic algorithm approaches for solving traffic signal coordination under oversaturation conditions. Comp.-Aided Civil and Infrastruct. Engineering 27(1): 14–28.

[96] Ramgulam, A., T. Erteken and P.B. Flemings. 2017. Utilization of artificial neural network in optimization of history matching. *In*: SPE paper 107468, Latin American and Caribbean Petroleum Engineering Conference, Buenos Aires, Argentina, 15–18 April 2007.

[97] Rauwolf, G.A. and V.L. Coverstone-Carroll. 1996. Near-optimal low-thrust orbit transfer generated by a genetic algorithms. Journal of Spacecraft and Rockets 33(6).

[98] Retes, M.P., G. Giorgi and J.V. Ringwood. 2015. A review of non-linear approaches for wave energy converter modelling. *In*: 11th European Wave and Tidal Energy Conference (EWTEC 2015), Proceedings of the 11th European Wave and Tidal Energy Conference, Nantes, France, September 2015.

[99] Retes, M., P. Mérigaud, J.-C. Gilloteaux and J.V. Ringwood. 2015. Nonlinear Froude-Krylov force modelling for two heaving wave energy point absorbers. *In*: 11th European Wave and Tidal Energy Conference (EWTEC 2015), Proceedings of the 11th European Wave and Tidal Energy Conference, Nantes, France, September 2015.

[100] Ringwood, J., G. Bacelli and F. Fusco. 2014. Energy-maximizing control of wave-energy converters: The development of control system technology to optimize their operation. Control Systems, IEEE 34(5): 30–55, October 2014.

[101] Rudolph, G. 1994. Convergence analysis of canonical genetic algorithm. IEEE Transaction on Neural Networks 5(1): 96–101.

[102] Saito, M. and J. Fan. 2000. Artificial neural network-based heuristic optimal traffic signal timing. Computer-Aided Civil and Infrastructure Engineering 15(4): 293–307.

[103] Scruggs, J.T., S.M. Lattanzio, A.A. Taflanidis and I.L. Cassidy. 2013. Optimal causal control of a wave energy converter in a random sea. Applied Ocean Research 42: 1–15.

[104] Soltani, M.N., M.T. Sichani and M. Mirzaei. 2014. Model predictive control of buoy type wave energy converter. *In*: The 19th International Federation of Automatic Control (IFAC) World Congress, Cape Town, South Africa, August 24–29 2014. International Federation of Automatic Control (IFAC).

[105] Song, J., O. Abdelkhalik, R. Robinette, G. Bacilli, D. Wilson and U. Korde. 2016. Multi-resonant feedback control of heave wave energy converters. Ocean Engineering 127: 269–278.

[106] Starr, B. 2012. Spooled DNA and hidden genes: The latest finding in how our dna is organized and read. The Tech Museum of Innovation, Department of Genetics, Stanford School of Medicine, 201 South Market Street San Jose, CA 95113.

[107] Sun, D.R., F. Benekohal and S.T. Waller. 2006. Bi-level programming formulation and heuristic solution approach for dynamic traffic signal optimization. Computer-Aided Civil and Infrastructure Engineering 21(5): 321–333.

[108] Suzuki, J. 1993. A markov chain analysis on a genetic algorithm. *In*: Proceedings of the Fifth International Conference on Genetic Algorithms, pp. 146–153.

[109] Suzuki, K. and N. Kikuchi. 1991. A homogenization method for shape and topology optimization. Computer Methods in Applied Mechanics and Engineering 93: 291–318.

[110] Teklu, F., A. Sumalee and D. Watling. 2007. A genetic algorithm approach for optimizing traffic control signals considering routing. Computer-Aided Civil and Infrastructure Engineering 22(1): 31–43.

[111] The National Institute of General Medical Sciences. 2010. The New Genetics. On line website, April 2010. National Institutes of Health.

[112] Vallado, D.A. 2004. Fundamentals of Astrodynamics and Applications. Microcosm Press and Kluwer Academic Publications, 2nd Edition.

[113] Varia, H.R. and S.L. Dhingra. 2004. Dynamic optimal traffic assignment and signal time optimization using genetic algorithms. Computer-Aided Civil and Infrastructure Engineering 19(4): 260–273.

[114] Vasile, M., E. Minisci and M. Locatelli. 2010. Analysis of some global optimization algorithms for space trajectory design. Journal of Spacecraft and Rockets 47(2): 334–344, March–April 2010.

[115] Vasile, M. and P.D. Pascale. 2006. Preliminary design of multiple gravity-assist trajectories. Journal of Spacecraft and Rockets 43(4): 794–805, July–August 2006.

[116] Venkataraman, P. 2009. Applied Optimization with MATLAB Programming. Wiley Publishing, 2nd Edition.

[117] Villegas, C. and H. van der Schaaf. 2011. Implementation of a pitch stability control for a wave energy converter. *In*: Proc. 10th Euro. Wave and Tidal Energy Conf. Southampton, UK.

[118] Vose, M.D. and G.E. Liepins. 1991. Punctuated equilibria in genetic search. Complex Systems 5: 31–44.

[119] Wall, B.J. and B.A. Conway. 2009. Genetic algorithms applied to the solution of hybrid optimal control problems in astrodynamics. Journal of Global Optimization 44(4): 493–508, August 2009.

[120] Wolgamot, H.A. and C.J. Fitzgerald. 2015. Nonlinear hydrodynamic and real fluid effects on wave energy converters. Proceedings of the Institution of Mechanical Engineers, Part A: Journal of Power and Energy 229(7): 772–794.

[121] Yang, L. 2011. Topology Optimization of Nanophotonic Devices. Phd, Technical University of Denmark, Department of Photonics Engineering, Building 343, DK-2800 Kongens Lyngby, Denmark.

[122] Yu, Z. and J. Falnes. 1995. State-space modelling of a vertical cylinder in heave. Applied Ocean Research 17(5): 265–275, October 1995.

[123] Zou, S., O. Abdelkhalik, R. Robinett, U. Korde, G. Bacilli, D. Wilson and R. Coe. 2017. Model predictive control of parametric excited pitch-surge modes in wave energy converters. International Journal of Marine Energy 19: 32–46, September 2017.

Index

A

Augmented Lagrange multiplier
method 65–67

B

Barrier method 61–63
Binary tags 97, 176
Buoy shape design 198

C

Chromosome 14, 79–81, 84–91, 94–97,
112, 113, 115, 117–125, 137, 141–143,
161, 162, 176, 182, 192
Conical buoy 187, 189
Conjugate gradient 52, 54, 56
Convergence rate 7
Crossover 77, 79, 80–82, 86, 87, 89, 93–98,
100–102, 108–115, 120–123, 140, 145,
161, 162, 172
Cyclic coordinated descent 48

D

Davidon-Fletcher-Powell Algorithm 57
Deep space maneuvers (DSM) 9, 19, 20,
23–25, 117, 121, 124, 126–134, 137,
140, 143–154, 56–169, 171–173,
175–180
Dependent variables 19, 20, 23, 24, 118,
121

DFP 57, 58, 66, 67
Direct optimization 11, 60, 68, 72, 77
DSMPGA 140–143, 145, 146, 149, 153,
186
Dynamic size multiple population
algorithms 137, 140, 141

E

Earth Mars mission 126–128, 137, 145,
146, 156, 164, 172
Earth Saturn mission 128–131
Equal interval 35–37, 40
Exterior penalty function 63–66, 70, 71

G

Global optimization 12–14, 75–78, 84, 97,
157
Golden section 37–40, 48, 53
Gradient-based optimization 51
Gram-Schmidt Orthogonalization 6, 7
Gravity assist model 21

H

Hidden genes binary tags 97
Hidden genes genetic algorithms
(HGGA) 76, 81, 84, 85, 86–89, 94,
97–101, 103–111, 115, 116, 137, 153,
160, 162, 164, 171, 172, 175, 176,
178–180, 186–188, 194, 200, 203
Hidden genes tags 97

I

Impulse response function 25–27, 183
Indirect optimization 60, 61, 63, 67
Interplanetary trajectory optimization 8, 19, 76, 100, 117, 126, 140, 153

J

Jupiter Europa orbiter mission 131

L

Linear programming 3, 42, 43, 68
Logic A 95, 96, 104, 105, 107, 114, 179, 180
Logic B 95, 104, 105, 107, 114
Logic C 95, 104, 105, 107, 115, 179, 181

M

Markov chain 80–82, 107, 109, 110
Mechanism A 95, 98–101, 104, 105, 107, 109–111, 177, 178, 180, 181
Mechanism B 95, 99–101, 105, 107, 111
Mechanism C 96, 99, 100, 102, 107, 108, 111, 112
Mechanism D 96, 99–101, 105, 107, 112
Mechanism E 96, 102, 103, 107, 112, 180
Mechanism F 96, 103, 107, 112, 181
Mechanism G 97, 103, 107, 113
Mechanism H 97, 103, 107, 112, 113
Messenger mission 149–154, 163
MGA 157
MGADSM model 127, 130, 131, 134, 166–169, 171–173, 178, 179, 181
Microgrid 11
Microgrid optimization 11
Multi-populations 118
Mutation 77, 79–82, 86, 89, 93–98, 100, 101, 105, 108–115, 120–123, 145, 172, 177

N

Niching 123, 124, 137–139
N-Impulse maneuver 19
Non-gradient optimization 51
Nonlinear programming 3, 41–45, 72, 73

O

Ocean wave energy converters 183
Optimal grouping 11, 12, 84

Optimization 2, 3, 7–16, 19, 25, 27, 30, 34, 37, 41–47, 49, 51, 53, 56, 60–64, 66–72, 74–79, 83–85, 87, 89, 90, 91, 95, 97, 100, 102, 107, 115, 117–119, 123, 124, 126, 128, 131, 137, 140, 142–146, 152, 153, 156–168, 172, 175, 176, 180, 183, 185–188, 191–193, 195, 197, 201, 203
Orbit design optimization 15
Orthogonal vectors 5, 6, 54
OSWEC 25, 26

P

Pattern search 48–51
Pixel classification 14, 84
Populations 14, 78–83, 85–89, 91–93, 98, 102, 108, 118–120, 123, 124, 127, 128, 131, 137–147, 149, 151, 154, 161, 164–167, 171, 172, 177, 181, 182, 191, 192
Powell 51, 57

Q

Quadratic programming 68, 69, 71–74

R

Radiation 25–27, 183, 187, 194
Rankine cycle 12, 13
Reproduction 77, 79, 80, 89, 92, 94, 140

S

Schema 80, 89–94
Sequential quadratic programming 68, 72, 74
Shape and control optimization 186, 197
Signal coordination 11
Space trajectory optimization 84, 119, 140, 142, 156, 157, 176
Steepest descent 52
Structural topology optimization 13
Structured chromosome differential evolution (SCDE) 117, 119, 120, 122, 123, 127, 128, 130–132, 135–137
Structured chromosome genetic algorithms (SCGA) 117–121, 123, 127, 128, 130–132, 135–137, 186
Success rate 98, 99, 101, 104–106, 126–128, 131, 132, 135, 136, 180

Systems architecture optimization 8, 75
Systems design 12

T

Tag 85, 94–102, 104–108, 110–115, 176,
177, 180–182
Tag evolution methods 104
Tag logical evolution 95, 104
Tag stochastic evolution 95
Taylor's theorem 4
Traffic network 11
Transformation in SCDE 122

W

Wave energy converter (WEC) 25–28,
183–185, 188, 194, 197, 200, 201
Wave surface elevation 25

Z

Zero-DSM Model 130, 134, 166–169,
171–173, 178, 179